14. 99

PRACTICAL GENETICS

PRACTICAL GENETICS

R. N. Jones
G. K. Rickards

Open University Press
Milton Keynes • Philadelphia

Open University Press
Celtic Court
22 Ballmoor
Buckingham MK18 1XT

and
1900 Frost Road, Suite 101
Bristol, PA 19007, USA

First Published 1991

British Library Cataloguing in Publication Data

Jones, R. N.
    Practical genetics.
    1. Educational institutions. Curriculum subjects. Genetics
    I. Title II. Rickards, G. K.
    575.10711

    ISBN 0–335–09218–7
    ISBN 0–335–09217–9 (pbk)

Library of Congress Cataloging-in-Publication Data

Jones, R. N. (Robert Neil), 1939–
        Practical genetics / R.N. Jones, G.K. Rickards.
        p. cm.
        Includes bibliographical references and index.
        ISBN 0–335–09217–9 (pbk.). – ISBN 0–335–09218–7 (hard)
        1. Genetics–Experiments. I. Rickards, G. K. (Geoffrey Keith),
    1939– . II. Title.
    QH440.4.J66 1990
    575.1'078–dc20
                                                                90–14233
                                                                CIP

Typeset by Stanford Desktop Publishing Services, Milton Keynes
Printed in Great Britain by St Edmundsbury Press, Bury St Edmunds, Suffolk

CONTENTS

PREFACE AND ACKNOWLEDGEMENTS

Practical work is an essential part of any introductory course in genetics. It is essential in order to communicate the subject matter and to stimulate interest and enthusiasm. If the subject is taught solely in a theoretical manner students can develop the impression that they understand the principles and the concepts involved, without actually doing so. In a practical situation, on the other hand, when faced with interpreting a set of results or observations students will quickly come to realise the depths and limitations of their understanding. They will appreciate, too, the gulf that exists between learning the theoretical details of, say, meiosis and actually making sense of the chromosome configurations they see down the microscope. Not only will they come face-to-face with the reality of the subject, but also they will be encouraged to think about it, and to develop a more critical attitude to what they read in textbooks. Apart from these considerations there is the added value of developing practical and technical skills for their own sake, and in knowing how to go about making observations and setting up exercises and experiments to demonstrate the principles and workings of heredity at first hand.

This book is about practical genetics. It is intended for the school/college/ university/polytechnic interface level for which there is a paucity of books that provide essential and relevant information. All of the material presented here is based on first-hand experience by the authors. The schedules and exercises are known to work repeatedly well, to have relevance to the theoretical basis of the subject, and to be within the means of a modestly equipped teaching laboratory. The organisms used commonly occur in many countries, in both northern and southern hemispheres. It is assumed that the reader is conversant with the theory of the subject and with the basic terminology of genetics, though we do provide a glossary for help in key areas. A special and important feature of the book is the careful and thorough treatment given to the analysis and interpretation of results and observations. It is relatively easy to carry out practical work according to a prescribed schedule, but it is an altogether more difficult task to use the information gathered in an instructive and beneficial way. We expect that the approach taken here will facilitate this latter part of the work. We have not tried to cover everything, because there is no definitive set of practical exercises. Neither have we attempted to be fashionable or technological for their own sakes. The aim has been to give good information and procedures about how to teach some basic aspects of genetics in a successful and stimulating way in a laboratory situation.

We alone are responsible for what is written. But we have been helped and guided by some of our colleagues in various ways. Some of these have read various chapters and have unselfishly offered advice and constructive criticism. Others have helped in general ways over many years to mould various exercises. In these respects, special thanks are due to Dr G.C. Hewitt, Dr J. Hutchinson-Brace, Dr M. Kearsey, Professor D.G. Lloyd, Mrs L. Milicich, Mr D.R. Romain, Dr H. Sealy-Lewis, Professor A.P. Wylie and Dr M. Young. In addition, Mrs M. Breese, Mrs K. Churchill, Mr D. Falding, Miss J.A. Silvester and Mrs C.J. Thorn all provided excellent technical assistance; and expert photographic work was done by Mr J.E. Casey, Mr G. Keating, Mr B. Robertson and Mr E. Wintle. A special word of thanks goes to Liz Haines for word processing. The British Council and the Royal Society of New Zealand generously provided financial support which enabled the project to get off the ground and proceed. We especially thank Mr Colin Ramsay of the British Council Wellington Office for his and the Council's continued support and enthusiastic encouragement.

R.N. Jones and G.K. Rickards

Chapter 1   *LOOKING AT CHROMOSOMES:MITOSIS*

The exercises described in this chapter are suitable for all levels of study, from senior school through to university undergraduate. All the plants and animals used are common species and they present no health hazards as far as is known. Suitable precautions, such as the wearing of laboratory coats and rubber gloves, should be taken when working with chemicals used for pretreatment, fixing and staining of chromosomes. These chemicals should not make routine contact with the skin and the fumes they give off should not be routinely inhaled. Feulgen's solution, which is used to stain chromosomes, is hazardous in that it colours the skin bright pink and can damage clothes and should therefore be handled with care. Similarly colchicine, which is used to pre-treat chromosomes, should be handled carefully to avoid it being absorbed through the skin.

INTRODUCTION AND BACKGROUND

Strictly speaking mitosis is division of the nucleus into two daughter nuclei that are identical to one another and to the parent nucleus. Mitosis is part of the cell cycle, which also includes chromosome replication and division of the cytoplasm (i.e. cytokinesis). However, the term mitosis is sometimes loosely used to include chromosome replication and cytokinesis with division of the nucleus itself. The chromosomes are always present in the nucleus, but they are most easily seen when the nucleus divides. During the division process the chromosomes contract considerably, after which they can be seen under the light microscope as distinct, thread-like structures. It is not usually possible to see chromosomes in living cells without using special methods of microscopy. The best we can do at this level is to obtain some tissue in which cells and their nuclei are actively dividing, fix the material to kill the cells and preserve their structure, stain the chromosomes with a dye and then look at them with a microscope after squashing the cells out flat.

By studying mitosis in this way we can observe the changes which take place in the chromosomes and see how these visible events bring about the process of heredity in somatic cells. Also, with the use of certain chemicals it is possible to arrest the division process at the end of the prophase stage and thus 'catch' the chromosomes in their most condensed and easily seen form. This technique permits the study of the number, size and form of the chromosomes, and enables us to see how these aspects of chromosome organization vary from one species to another.

PROCEDURES

Conventions for Writing Chromosome Numbers

The conventions for writing chromosome numbers are as follows:*

n = the number of chromosomes in the gametes, i.e. the *gametic number*;
$2n$ = the number of chromosomes in the zygote, i.e. the *zygotic number*; and
x = the *basic number*, i.e. the number of different chromosomes in a haploid set.

In a diploid the basic number is the same as the gametic number and therefore $x = n$. We therefore write the chromosome number of the onion, for example, as $2n = 2x = 16$. The $2n$ tells us that we are giving the zygotic number of chromosomes and thus the number of chromosomes present in somatic cells of the organism. The $2x$ tells us that the species is diploid in that it has two sets of $x = 8$ chromosomes. We do not simply use $2n = 16$ because many species of plants, and some animals, are not diploids: they are polyploids, having various multiples of the basic set of chromosomes.

To denote the level of ploidy we simply give the appropriate value of x. Thus for the polyploid series of the hyacinth, which is based on $x = 8$, we write:

$2n = 2x = 16$,   diploid;
$2n = 3x = 24$,   triploid;
$2n = 4x = 32$,   tetraploid, etc.

It is incorrect to use $3n$, $4n$, etc. to represent these different levels of ploidy: $3n$ and $4n$ strictly have no meaning. Where the ploidy status of a species is unknown we have to use just the $2n$ value, such as in $2n = 42$. To confirm the ploidy status it is usually necessary to examine chromosome pairing behaviour in meiosis.

In animals polyploidy is uncommon and it is the custom, therefore, to give the chromosome number simply as $2n = 46$ (human), $2n = 8$ (Drosophila), for example, and to take it for granted that we are dealing with a diploid. But when polyploidy is known we give all the information we have, such as in the frog *Hyla crysoscelis* ($2n = 2x = 24$) and its presumed tetraploid derivative *Hyla versicolor* ($2n = 4x = 48$).

General Procedures for Making Slides to Study Mitosis

Material to use

Cells dividing by mitosis can be found in the actively growing parts of an organism and in tissues where the cells require continual replacement. In plants it is usual to work with meristematic regions of the roots, although shoot tips can also be used. With animals the source of material may be young embryos or spermatogonial cells from the testes of male insects such as locusts and grasshoppers. In more advanced work using mammals (including humans), mitosis can be studied in white blood cells grown in a culture medium.

For most teaching purposes mitosis is best studied in the meristematic regions

---

* Note that in some publications capital rather than lower case letters are used.

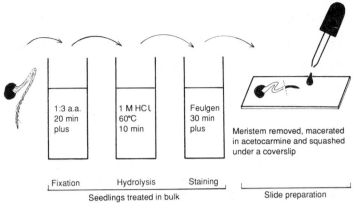

Fig. 1.1 Sequence of operations in the preparation of slides for the study of mitosis in onion seedlings (a. a. = acetic alcohol).

of plant roots (the 'root tips'). These can be produced by germinating seed on moist filter paper or in vermiculite. Alternatively, root tips can be sprouted from bulbs or corms suspended over water. Many monocotyledonous plants are ideal material to use because their chromosomes are often large and few in number and their roots easy to process. *Tulbaghia violacea* , for example, has 2n = 2x = 12 with very large chromosomes. Plants of this species can be obtained readily as corms for easy growth and maintenance in pots. In the autumn there is also a wide choice or ornamental flower bulbs such as hyacinth, crocus and grape hyacinth, all of which give good results using roots that will form when bulbs are placed in contact with water. For pot plants such as *Tradescantia* and *Tulbaghia* it is often possible to find actively growing roots protruding through the drainage holes at the base of the pot, or to get roots by up-ending and removing the pot to expose the root system.

Allium cepa, the bulb onion, is an ideal material to use in almost all respects. Other diploid species of *Allium* also give very good results. The broad bean, *Vicia faba*, is especially good for large classes because root tips of this species can be produced in abundance and easily stained. Details of how to obtain and process root tips of these two species in particular are given as specific exercises in a later section. For mitosis in human cells, see Chapter 9.

There are three main stages to the general procedure for preparing material for the study of mitosis (Fig. 1.1), namely fixing, hydrolysis and staining.

Fixing material

Root tips or other tissues to be used for a study of mitosis are first put into a *fixative*. The most widely used fixative is a *freshly* prepared mixture of one part of glacial acetic acid and three parts of absolute ethyl alcohol (1 : 3 acetic alcohol); see Appendix for details concerning preparation of solutions. The fixative kills the tissues and preserves them with the minimum of damage and distortion to the cell contents. Fixation takes at least 30 min, but when a large quantity of material (for example, hundreds of seedlings) is being processed at the same time it is advisable to allow a longer time in the fixative and to change the fixative a few times over the first hour. Once fixed the material can be kept for a few days at room temperature. It is nearly always best to use recently fixed material, but if prolonged storage (several weeks) is necessary the material should be transferred from the fixative to 70% ethyl alcohol and kept in the refrigerator at 2–5°C, otherwise the tissue will become hard and brittle. Bottles with stored material should be well sealed to prevent evaporation of the alcohol.

Hydrolysis

Most procedures require fixed material to be *hydrolysed*, by treatment with hydrochloric acid (HCl). Hydrolysis brings about certain changes in the DNA of the chromosomes and allows for the stain to react specifically with the genetic material and not at all or just weakly with other components of the cell. It also softens the material, making it easy to macerate into fine pieces.

For the Feulgen staining procedure (see below) material is hydrolysed by immersing it in 1 M HCl at 60°C for 10 min, during which both temperature and time are important. The acid, therefore, must be *preheated* in tubes in a water bath or incubator, to ensure that the full 10 min at the correct temperature is given, otherwise the stain will not take well. Also, if a lot of material is to be hydrolysed for Feulgen staining it should be well drained of fixative (on to blotting paper) to ensure that undue dilution and cooling of the HCl does not occur. When the required 10 min hydrolysis is up, transfer the material quickly to the stain (see below), or if there is a time lag in doing this add water to the acid to cool and dilute it.

For carmine or orcein staining, hydrolysis is normally achieved simply by mixing a few drops of 1 M HCl with the stain and then heating the two together. Satisfactory hydrolysis of some material such as broad bean root tips can be achieved by treatment with 5 M HCl for 45 min at room temperature.

Staining

Many dyes are available for staining chromosomes to make them clearly visible under the microscope. We recommend the use of *basic fuchsin* in the form of *Feulgen's solution* (Schiff's Reagent) since when the technique is carried out properly the chromosomes will be stained a brilliant red-purple against a colourless background. After hydrolysis the softened material is gently drained of surplus liquid and transferred to the stain at room temperature. After 30 min the meristematic regions of roots, for example, will be coloured strongly, indicating that the nuclei and chromosomes have taken up the stain. If there is any delay in making slides after staining with Feulgen's it is best to leave the material in the stain in the dark, ideally in a refrigerator.

Carmine or orcein dyes dissolved in acetic acid or propionic acid are also good stains to use when making slides of mitosis. They are applicable to a wide range of plant and animal materials and are good to use if any difficulties arise with the Feulgen staining technique. Material is transferred to the dye in a watch glass and heated over a spirit lamp or low flame of a bunsen burner. While doing this the watch glass should be held firmly with a pair of stout tweezers or placed on a tripod. Alternatively, the material can be heated directly on a microscope slide. The heating drives the stain into the chromosomes, though for some material, such as the broad bean, staining can be achieved without heating.

Slide preparation

After staining, material is placed on a glass microscope slide and macerated in a small drop of 45% acetic acid or in a drop of aceto-carmine or propionic ocein stain (Fig.1.1). Make sure the material does not dry out on the slide before adding liquid to it. The acetic acid serves as a medium in which to macerate the material. Aceto-carmine or propionic orcein serves the same purpose and may also enhance the staining of the chromosomes and give some counter stain to the cyto-plasm. The amount of liquid to use will be judged by experience. It often helps in the subsequent maceration step to start off with a bare minimum of liquid, just sufficient to keep the material moist. Further liquid is then added after macera-tion. The final quantity of liquid should not exceed the amount that can be accommodated under the coverslip before squashing.

Macerate the tissue into fine pieces with a flat bottomed glass or aluminium rod. This part of the procedure should be done carefully, otherwise the subse-quent squashing step will fail. Resist the temptation to try and macerate too much material on the same slide. The idea is to separate out small pieces of tissue and individual cells so that they form a single layer under the coverslip. Remove any pieces which cannot be macerated by picking them out with a needle or a pair of fine forceps.

After thorough maceration, distribute the material and apply a coverslip. Then, gently to begin with, push down the coverslip between two pieces of blotting paper. Finally, place the thumb over the blotting paper and coverslip and exert pressure, which should be applied directly downwards to avoid smearing. Additional flattening and spreading of cells can be achieved by tapping on the coverslip with the glass or aluminium rod as a follow-up to the thumb treatment. It may help to warm the slide slightly over a low bunsen flame or spirit lamp.

Air spaces can be filled by allowing a little acetic acid or stain to run under the edges of the coverslip, followed by blotting off excess liquid. Rubber solution or nail varnish can be used to seal the coverslip so as to make a temporary mount, which will keep for at least a day, especially if kept in a moistened chamber such as a petri dish. The procedure for making slides permanent is given in the Appendix.

EXERCISES

1 Making Slides of Mitosis using the Onion, *Allium cepa*

Obtaining root tips

From germinating onion seed
Roots can easily be obtained from onion seedlings grown on moist filter paper in a petri dish at any time of the year. The garden variety 'White Lisbon' is particu-larly suitable due to its rapid rate of growth. To obtain roots, proceed as follows:

1 Place a double thickness of filter paper in each of several petri dishes and moisten it with distilled water. Drain off any surplus water and then sprinkle about 50 seeds onto the prepared dishes. Keep the dishes covered at room tem-perature and water daily. Any seeds that show signs of fungal infection should be picked off and discarded.

2 After 3 days growth transfer the germinating seeds to fresh dishes and space them out. It is important to keep the roots growing well and to avoid any impediments due to fungal diseases, overcrowding or 'drowning' with water. The seedlings will be ready for use after about 5 days of growth and can be kept in a good condition for a few days longer.

From onion bulbs

Onion roots can also be obtained easily from bulbs. To do this, use bulbs that have been tested beforehand for good root production, because commercially produced bulbs are often treated to prolong their dormancy. Shallots are seldom treated in this manner and overall are the best variety to use.

1 With a razor blade, thinly shave the base of the bulb to expose new tissue. Remove dry leaves from the bulb.
2 Place the bulb over the mouth of a glass container nearly filled with fresh water so that the base of the bulb just sits in the water. For wide-neck containers or small bulbs, use stainless steel or aluminium pins or probes to keep the bulb suspended and in place.
3 Keep the container topped up with water and change the water completely every 2 days or so. Roots will appear in 3 – 5 days.

Fixing material

Fix onion root tips in freshly prepared acetic alcohol (see Procedures). For roots grown from bulbs, cut off and place only the root tips (1–2 cm lengths for ease of handling) into fresh acetic alcohol. In either case it is best to fix material around mid-day, when the mitotic rate is at its peak.

Hydrolysis, staining and slide preparation

Hydrolyse and stain material according to the schedule given in the Procedures section, using either Feulgen's or propionic orcein stain. For *staining with Feulgen's*, exercise special care with the hydrolysis step (precisely 10 min at 60°C). After hydrolysis, transfer the material to the stain carefully for it will be soft and thus easily broken. Allow at least 30 min for the stain to take. After staining, transfer material to a slide, cut off the heavily stained meristem, and discard unwanted material. Macerate material in 45% acetic acid or propionic orcein. Add a coverslip and squash. Ring the coverslip with rubber solution or nail varnish.

Alternatively, stain material with propionic orcein. To do this, transfer a few root tips to a watch glass. Add a small quantity of propionic orcein (or aceto-carmine) stain plus a drop or two of 1 M HCl (approximately 10 parts of stain to 1 part of HCl). Heat gently (but do not boil) for $\frac{1}{2}$ – 1 min. Transfer the stained material to a microscope slide. Cut off the meristem and discard the remainder. Add a drop of fresh stain. Macerate and complete slide preparation as above.

2 Making Slides of Mitosis using the Broad Bean, *Vicia faba*

Vicia faba is also good material to use in a study of mitosis since its root tips are easy to produce and process in abundance. For junior classes or large classes, fixation and staining of material can be done beforehand by a teacher or technician, leaving the final and very easy staining step for individual students to do.

Obtaining root tips
First soak *fresh* seeds in water overnight. Then plant them in well moistened vermiculite for 5–7 days (5 days at 25°C in continuous light is ideal). When the radicles are about 2–3 cm in length pinch off the terminal meristems and also pinch off the shoot tips. Replant the seedlings for a further 5–7 days to obtain a good supply of lateral roots (again use continuous light if possible).

Fixing material
Remove seedlings from the vermiculite and wash them thoroughly. Cut 1–2 cm length root tips into freshly prepared acetic alcohol and leave for at least 1 h. If storage is necessary, transfer the material to 70% alcohol.

Hydrolysis, staining and slide preparation
Hydrolyse material with 5 M HCl at room temperature for 45 min. Then wash the root tips in water and, if necessary, keep them in cold water (4°C) for up to 2 weeks. For making slides, place a single root tip on a clean slide, identify the meristematic end, cut it off and remove unwanted material. Place a small drop of propionic orcein over the meristem. Macerate with a tapper and then apply the coverslip and squash according to standard techniques (see Procedures). The stain takes instantaneously. Ring the slides with rubber solution or nail varnish to make temporary mounts.

Broad bean roots can also be stained by the Feulgen method, using the procedure outlined for onion roots.

3 Studying and Understanding Mitosis from Prepared Slides

For studying and understanding mitosis in a practical class it is essential that students look at good quality slides. These can either be made by students themselves, as described above, or ready-made slides can be provided. When making their own slides, it is important to encourage students to make *a number* of slides, not just one, before proceeding. In this way, slide making technique is allowed to improve; and students often are unaware just how very good their slides can be. As a back-up, it is useful to have available spare slides made previously, either as temporary preparations made just before class or as permanent preparations on which areas of good cells are marked. Some laboratory situations may allow only a demonstration of slide making technique, followed by students studying slides already made by a technician or teacher or purchased from a biological supply house.

What we can see when we look at a slide of mitosis down the microscope will depend not only on the quality of the preparation but also on the equipment that is used. Some notes on microscopy are given in the Appendix. A x40 objective is the minimum requirement to see anything worthwhile, and a x63 or x100 oil immersion lens is usually needed to resolve the doubleness of the chromosomes at prophase and to see details of orientation at metaphase. It is highly desirable to have at least one microscope in the laboratory which is equipped to this high standard.

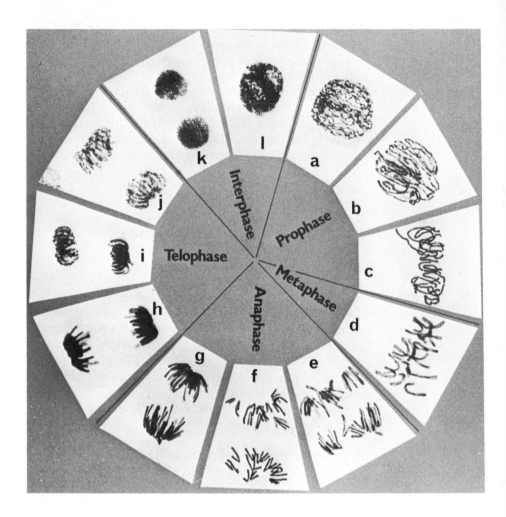

Fig. 1.2   Mitosis and interphase nuclei in the onion (*Allium cepa*, **2n = 2x = 16**).

Photographs of mitosis in *Allium cepa* and *Tulbaghia violacea* are shown in Figs 1.2 and 1.3a, with interpretative drawings of the latter in the accompanying Fig. 1.3b. The photographs for these figures were taken at a magnification of x630 onto a 35 mm film, using a x63 oil immersion objective and a x10 lens in the camera tube. For comparable photographs of mitosis in human cells, see Chapter 9.

When looking at slides of mitosis, variation in the size of interphase nuclei can be pointed out and discussed in relation to the synthesis of DNA and chromosome replication. Nucleoli will be seen as unstained 'ghosts'. Some odd-shaped nuclei that are associated with differentiated cells will also be seen. In very favourable cells it may be possible to confirm the chromosome number by counting at metaphase or anaphase, or to study the morphology of the various chromosomes in the set.

The important approach when looking at mitosis from slides is to emphasize the *events* that are taking place rather than simply to identify the named stages.

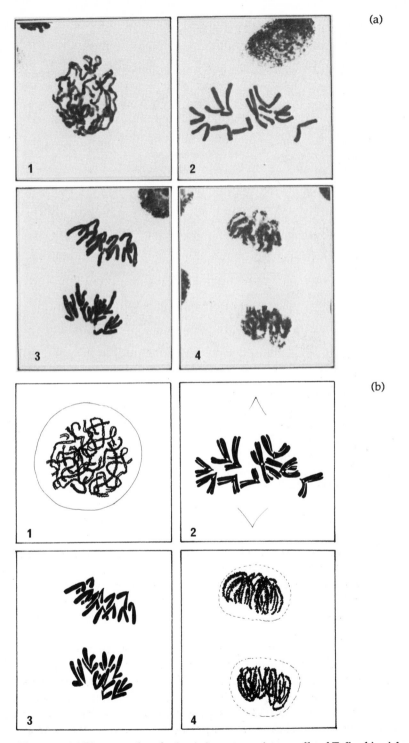

Fig. 1.3   (a) Photographs of mitosis in root meristem cells of *Tulbaghia violaceae* (2n = 2x = 12). (b) Interpretative drawings of the chromosomes shown in the photographs in (a).

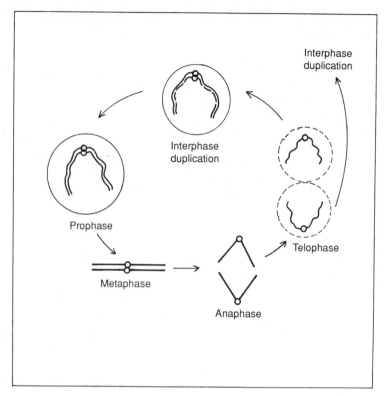

Fig. 1.4   Diagram showing the essential features of chromosome replication and mitosis.

Students should look critically at a range of cells and ask such questions as: (i) are chromosomes clearly visible as discrete structures? (ii) how well contracted are the chromosomes and are chromatids visible? (iii) are the chromosomes organized (oriented) in any particular way? (iv) have the chromatids separated into two groups? After studying a range of cells students should be encouraged to draw *actual* cells as they appear, and then present interpretative diagrams. The drawings need not include all of the chromosomes. The interpretation and the drawings of the various events of mitosis will be made more meaningful if pupils consider, as part of their practical work, the essential features of mitosis as outlined in Fig. 1.4 and summarized as follows.

Contraction of chromosomes : prophase
Observations of prophase cells will reveal that chromosomes gradually become shorter and thicker (by 'contraction' or 'condensation') and thus become more and more obvious (Fig. 1. 2 a,b). Two or three stages of this contraction should be recorded by students in their drawings. They might then consider the possibility that the *role* of contraction in prophase is to allow for movement at later stages of mitosis without chromosomes becoming entangled or broken.

Careful observation of prophase (and metaphase) cells under high magnification will verify the fact that each chromosome body is a *double* structure, being composed lengthwise of a pair of *sister chromatids*. These sister chromatids are going to separate from each other and move to opposite ends of the cell during anaphase.

Orientation of chromosomes : metaphase

Cells in which the chromosomes have moved (congressed) into the equator of the cell will be easily found and should be studied carefully (Fig. 1.2c, d). In this orientation stage of mitosis the chromosomes become attached at their centromeres to the spindle, such that sister chromatid centromeres are attached to opposite poles. The spindle will not be visible in cells prepared according to the techniques described above, but the centromeres can be located lying along the equator. The chromosome arms, therefore, lie haphazardly towards the poles (Fig. 1.2d). Orientation is a crucial event of mitosis and thus should receive special attention.

Separation of chromatids : anaphase

Sister chromatids are held together by a cohesive force (of unknown nature) up until the beginning of anaphase. At the beginning of anaphase this cohesive force lapses to allow the chromatids to move to opposite poles of the spindle. This event can easily be studied (Fig. 1.2e–h). In these anaphase cells the approximate positions of the centromeres can be determined because they now point directly to the poles, whilst the chromosome arms trail behind (see especially Fig. 1.2f).

Reconstruction of the interphase nucleus : telophase

Stages showing the reformation of the interphase nucleus during telophase (Fig. 1.2i–k) are a little difficult to find, because the chromosomes have lost much of their distinctiveness. By closing the iris diaphragm of the microscope it may be possible in some of these telophase cells to find evidence of cytokinesis (division of the cytoplasm) having taken place.

Chromosome replication for mitosis and its outcome

After studying mitosis itself students should be encouraged to consider the *origin* of each pair of sister chromatids, i.e. to consider DNA and chromosome replication during interphase. This event cannot be seen directly and so is illustrated diagrammatically in Fig. 1.5. This diagram will also help in understanding the *outcome* of mitosis, i.e. the production of cells with one and the same chromosome number and genetic composition.

4 C-mitosis and Karyotypes

More interest and additional scope for practical work can be given to the study of mitosis if *pre-treatments* are used to disrupt the organization of the spindle so as to block the mitotic cycle. When this is done the chromosomes can be seen much more clearly, because they are not situated at the equator of the spindle in their normal metaphase arrangement (Fig. 1.6). The most effective way to induce this condition is to use a drug called *colchicine* - hence the term colchicine or *c-mitosis*. Procedures for obtaining root meristems, and for staining and the preparation of slides, are the same as described previously. The only difference in the schedule

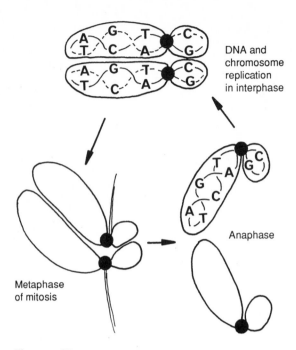

DNA and
chromosome
replication
in interphase

Anaphase

Metaphase
of mitosis

Fig. 1.5 Illustration of DNA and chromosome replication for mitosis. An anaphase chromosome is shown containing a single duplex of DNA with four base pairs (two each of A:T and G:C). During replication the two strands of 'original' DNA (solid lines) separate from each other into sister chromatids, each strand having produced on itself a complementary strand of 'new' DNA (dashed line, top diagram). Therefore the sister chromatids are genetically the same. Also shown is metaphase (with spindle fibres attached to centromeres) and anaphase.

is that the seedlings (or excised roots) are totally immersed in an aqueous solution of colchicine for 2–6 h *before* fixation. The concentration of colchicine to use is 0.1% (w/v). The optimum duration of treatment varies from one species to another: it is best determined by experience and experimentation. *Allium cepa* and *Vicia faba* give good results when treated for about 4 h at room temperature. During the colchicine pre-treatment, nuclei approaching metaphase will be arrested in their division cycle, because there is no spindle, and will accumulate at this stage. The chromosomes will also be more contracted than in a normal metaphase, sometimes with their chromatids spread apart except at their centromeres, due to a loss of chromatid cohesion.

There are various other spindle-inhibiting substances that can be used in the same way as colchicine. By and large they work just as well and they are much less expensive (see Appendix for details of preparation). The commonly used ones are para-dichlorobenzene (2–6h), α-bromonaphthalene (2–6h), 8-hydroxyquinolene (2–6h) and cold water (24h). Bromonaphthalene is recommended for use in schools where mitosis practical classes may be run only once a year and where colchicine is not routinely available. Photographs of c-mitosis in several species of plants are shown in Fig. 1.6. Compare these with the normal metaphase in Fig. 1.2.

Pre-treatments of the kind described above are very useful for the purposes of

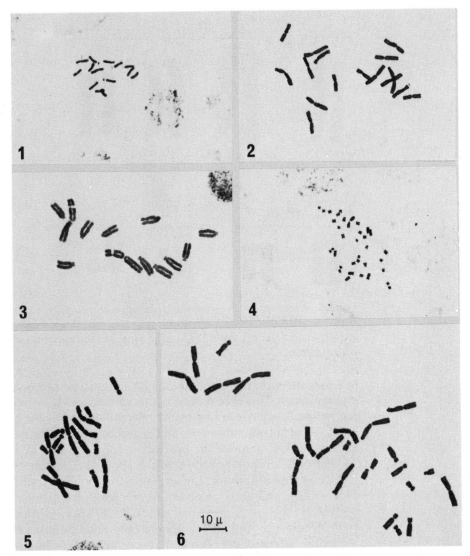

Fig. 1.6 C-mitosis in several species of plants, all at the same magnification (*c.* x 2000). Note the variation in the number, form and size of the chromosomes in the various complements. (1) *Pisum sativum*, **2n = 2x = 14**; (2) *Allium cepa*, **2n = 2x = 16**; (3) *Vicia faba*, **2n = 2x = 12**; (4) *Muscari armeniacum*, **2n = 4x = 36**; (5) *Hyacinthus orientalis*, **2n = 2x = 16**; (6) *Hyacinthus orientalis*, **2n = 3x = 24**.

counting chromosomes and for constructing *karyotypes*. A karyotype is made by cutting out individual chromosomes of a set from a photograph and then rearranging and classifying them according to their form (i.e. the position of their centromeres and nucleolus organizers) and relative lengths. For some species there are standard published karyotypes in which the chromosomes are numbered and identified on an internationally agreed basis, as in humans. These standard karyotypes are used for reference when screening and describing chromosome mutations, and for studying the relationships between different species.

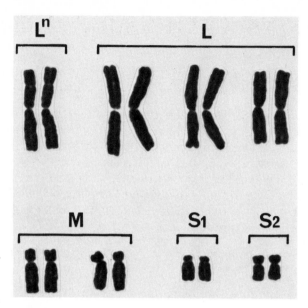

Fig. 1.7 Karyotype of the hyacinth (*Hyacinthus orientalis*, **2n** = **2x** = 16). The chromosomes have been classified into four long pairs (L), two medium-sized pairs (M) and two short pairs (S1 and S2). Ln denotes the long pair that has the nucleolus organizers (secondary constrictions).

As an alternative to photography, the karyotype can be made by drawing the chromosomes. This is best done with a camera lucida, or with the aid of a simple micrometer eyepiece to obtain the relative lengths of the chromosomes and the positions of their centromeres and secondary constrictions (nucleolus organizer regions). A karyotype of the hyacinth is shown in Fig. 1.7 and that of *Allium triquetrum* (a relative of the bulb onion) is shown in Chapter 2 (Fig. 2.5b).

A good way of approaching an exercise involving karyotypes is for students to make slides and then compare chromosome complements of two (or more) species of a genus. A good pair of species to study is *Allium cepa* and *A. triquetrum* . These two diploid species not only have different chromosome numbers (16 vs 18) but also have clear differences in chromosome sizes, centromere positions and numbers of nucleolus organizer regions (Fig.1.6 and Chapter 2 Fig. 2.5b).

REFERENCES

Darlington C.D. and LaCour L.F. (1976). *The Handling of Chromosomes*, 6th edition. London, Allen and Unwin.

Dyer A.F. (1979). *Investigating Chromosomes*. London, Edward Arnold.

John B. and Lewis K.R. (1972). *Somatic Cell Division. Oxford Biology Reader*, eds J.J. Head and O.E. Lowenstein. Oxford, Oxford University Press.

Jones R.N. and Karp A. (1988). *Introducing Genetics*. London, John Murray.

MacGregor H. and Varley J. (1988). *Working with Animal Chromosomes*, 2nd edition. New York, John Wiley and Sons.

McLeish J. and Snoad B. (1972). *Looking at Chromosomes*. Basingstoke, Macmillan.

Chapter 2 LOOKING AT CHROMOSOMES: MEIOSIS

The work in this chapter is suitable for all levels of study from senior school through to university undergraduate. The plants and animals used present no hazards. Careful attention should be given to conservation issues if material is collected from natural populations. Good standards of laboratory practice, such as wearing of gloves and a laboratory coat, should be ensured when handling chemicals used for fixation and staining. The chemicals used are not dangerous, although they should not routinely make contact with the skin; and the fumes they give off should not be routinely inhaled.

INTRODUCTION AND BACKGROUND

Meiosis consists of two special divisions which together reduce the zygotic (2n) number of chromosomes to the gametic (n) number. This reduction compensates for chromosome doubling which occurs when gametes (sperm and egg) fuse at fertilization. In diploid organisms, meiosis produces haploid cells each of which contains only one of each type of chromosome, whereas the diploid zygote and body cells contain pairs of such chromosomes. Also, the cells produced by meiosis contain combinations of alleles different from those that were present in the parents of the organism concerned.

In this chapter we summarize the major events of meiosis in a diploid organism and then describe how meiosis can be studied in onions and other plants and in insects. The exercises we present are designed to show how a good understanding of meiosis can be achieved through study of its important features from prepared slides and photographs.

Fig. 2.1 shows a plan of meiosis and introduces some of the terms we will be using. Fig. 2.2 illustrates diagrammatically the main events of meiosis, which are summarized as follows.

1 Prior to the initiation of meiosis, each chromosome *replicates* and thus *chromatids* are produced.
2 Homologous chromosomes *synapse* (align with each other closely all along their lengths), thereby producing the haploid number of *bivalents*.
3 Within each bivalent at least one *chiasma* forms (by breakage of chromatids followed by their rejoining in new combinations).
4 Chromosomes *contract* by coiling; and homologous chromosomes separate from each other somewhat, except at their chiasmata.

The above events take place during *prophase* of the first division of meiosis (i.e. prophase I), which is often sub-divided into *leptotene* (prior to pairing), *zygotene* (pairing), *pachytene* (pairing is complete), *diplotene* (pairing lapses) and *diakinesis* (opening out of chiasmata and completion of contraction).

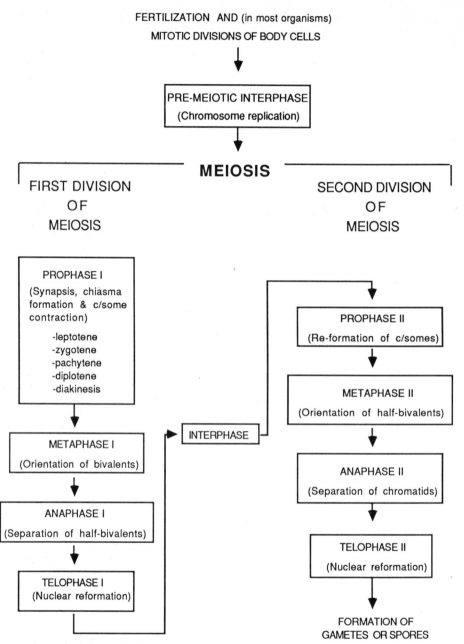

Fig. 2.1   A plan of meiosis, showing key events and stages.

5 The nuclear envelope disappears. Each bivalent becomes connected at its centromeres to opposite poles of the spindle and moves to the centre (equator) of
   the spindle. We use the term *prometaphase* for the period during which these

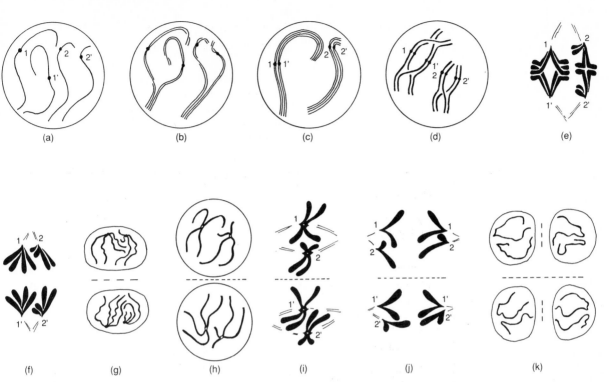

Fig. 2.2 Diagram of the main events of meiosis in a diploid organism with two pairs of chromosomes, labelled at their centromeres 1, 1′ (a metacentric pair of chromosomes) and 2, 2′ (an acrocentric pair). The cytoplasm is not shown. (a) End of the preceding mitosis, prior to chromosome replication. (b) Beginning of synapsis, after chromosome replication; prophase I (zygotene). (c) Completion of synapsis (bivalents fully formed); prophase I (pachytene). (d) Homologues have separated from each other except at chiasmata; prophase I (diplotene.) (e) Bivalents have oriented along the equator and have spindle fibres connecting the centromeres to the poles; metaphase I. (f) Separation of half-bivalents; anaphase I. (g) Reformation of nuclei and cytokinesis (dashed line); telophase I and end of the first division of meiosis. Each nucleus proceeds separately (though often synchronously) through the second division. (h) Reformation of chromosomes; prophase II. (i) Orientation of half-bivalents; metaphase II. The second division spindles are at right angles to that of the first division. (j) Separation of chromatids; anaphase II. (k) Reformation of nuclei and cytokinesis; telophase II and end of meiosis.

movements take place; and *metaphase* when the movements have finished.

6   Bivalents separate completely into *half-bivalents*, which move to opposite poles of the spindle *(anaphase I)*.

7   After reformation of nuclei and division of the cytoplasm *(telophase I)* an *interphase* stage usually occurs.

8   Chromosomes contract a second time *(prophase II)*; and chromatids become connected to opposite poles of a new spindle *(metaphase II)*.

9   Chromatids (chromosomes) move to opposite spindle poles *(anaphase II)* and new nuclei form *(telophase II)*.

PROCEDURES

Where to Find Meiosis

Meiosis will be found in the testes and ovaries of an animal when it is producing sperm and eggs; and in plants it will be found in the anthers and ovaries during the production of pollen and ovules. Generally it is easiest to study male meiosis in the anthers of plants. However, some insects provide very good material for the study of meiosis, especially of chiasmata. A combined study of meiosis in pollen mother cells in a plant and of chiasmata in spermatocytes of an insect is a very good way of understanding the events of the meiotic process in a laboratory situation.

Material to Use

In selecting a suitable material it is important that its chromosome number is low ($n = 5$ or 7 is ideal) and that it has a range of chromosome sizes, centromere positions and chiasma characteristics. Material should be able to be grown or collected from natural populations (with proper consideration of conservation issues) in abundance, and when fixed should be able to be stored indefinitely without deterioration. Flowers of the appropriate age should be easy to dissect under a microscope; and should contain numerous anthers at different ages, to enable all of the main events of meiosis to be seen without too much effort.

Members of the Liliaceae and especially those of the onion genus *Allium* (of which there are many members) are ideal for studying meiosis in almost all respects. Rye *(Secale cereale)* is also very good, as is *Pinus radiata* and paeonies.

Consult a local botanist if you intend to use material collected from natural populations, to be sure that you are not dealing with an endangered or potentially endangered species.

General Procedures for Making Slides to Study Meiosis in Flower Material

We will use *Allium triquetrum* (garlic onion, wild garlic or onion weed; Fig. 2.3a) to describe the procedures to use. But these procedures are applicable to any member of the onion genus, and with little or no modification are applicable also to a wide range of other plants including rye and other cereal grasses, members of the buttercup family (Ranunculaceae) such as paeonies, and even male cones of *Pinus radiata*.

Collecting material

Allium triquetrum is found in many countries, often as a weed growing in moist, shaded areas. It can be cultivated easily from bulbs. Onion plants can also be produced in abundance from seed (but note that these may require two seasons' growth before producing flowers).

Meiosis in onions has finished by the time the flowers are conspicuous (Fig. 2.3a); all you will see in anthers of such flowers will be pollen grains (Fig. 2.3e). It is necessary, therefore, to collect young flowers, well before they have become conspicuous. To do this for *Allium triquetrum* it is best to collect whole

Fig. 2.3 *Allium triquetrum.* (a) Mature plant. The flowers of the two showy inflorescences will have long since completed meiosis. (b) A young inflorescence (white arrow) about to be removed with tweezers from the base of the plant, after peeling down its leaves. (c) Removing the protective bracts (middle and lower arrows), prior to putting the inflorescence into fixative (in bottle at right). The uppermost arrow indicates the size of flower in which meiosis will most likely be found. (d) Pollen mother cells undergoing meiosis. The arrows locate non-meiotic cells of the tapetum (nutritive tissue). (e) Pollen grains (arrows) after they have been liberated from the cell wall in which meiosis had taken place. At upper right a portion of the anther wall is shown. Nuclei of two tapetal cells lie amongst the pollen grains.

plants by uprooting them, washing off soil and then placing them in a damp plastic bag for transport to the laboratory. The young flowers are grouped into inflorescences, which will be found by peeling the leaves down towards the bulb (Fig. 2.3b). Large (fat) plants will usually have more than one inflorescence at the desired age. With a little practice no trouble will be experienced in finding the right inflorescences to collect. They will still be white or just starting to turn green. Snip off each inflorescence with fingers or tweezers. The flowers will be covered by two protective bracts, which should be removed so as to hasten fixing and staining (Fig. 2.3c).

Fixing and storing material

Place the inflorescences immediately into fixative (meiosis is very sensitive to moisture loss), using freshly prepared acetic alcohol (see Appendix). Material should be left in the fixative for 24 h and then transferred to 70% alcohol and stored if desired in a refrigerator. Material stored for 5 years (at least) is still in good condition, though containers should be well sealed to prevent evaporation of the alcohol.

Staining material

For screening material to find flowers of the right age, fixed material (and even fresh material) can be dissected and stained in that order, in which case aceto-carmine or aceto-orcein or, better still, propionic orcein (see Appendix) should be used as a stain. In this technique, anthers are removed from a flower (see below for details) and the anther contents squeezed out into a drop of stain on a clean slide. After removing the anther wall and applying a coverslip the slide is gently heated over a small bunsen flame or spirit lamp for a minute or so to accentuate the stain.

For most class use, however, it is best to stain material in bulk with Snow's alcoholic carmine (see Appendix) *before* dissection. To do this, transfer appropriate quantities of fixed material to the stain and leave it at room temperature for 72 h. Material left in the stain keeps well for at least 1 month; longer if it is kept in a refrigerator. Immediately prior to use, transfer the material to a watch glass or petri dish and cover it with 70% alcohol acidified to 10% with HCl. Cover the material to prevent evaporation of the alcohol.

Dissecting material and slide preparation

Making preparations from whole flowers leads to difficulty in identifying meiotic cells amongst a mass of developing flower material, some cells of which will be undergoing mitosis! Thus the best starting procedure is to dissect the flower and use whole anthers as follows:

1 With a scalpel, cut the flowers from the inflorescence stalk and arrange them according to size. Select a flower of appropriate size (determined by trial and error) and place it on a slide, taking mental note of its size in relation to others.
2 Using a dissecting microscope, hold the flower at its base with tweezers and remove the anthers with a probe. Distribute the anthers over the approximate area of a coverslip and discard remaining flower material. Apply a coverslip (which will hold the anthers in place). Run a little 45% acetic acid or propionic-orcein under the coverslip.

3 Apply firm thumb pressure to the coverslip between folded blotting paper to burst the anthers and liberate the pollen mother cells, which will remain as a halo surrounding the flattened anther wall.

After some experience you may wish to burst the anthers with a blunt instrument *before* the coverslip is applied. Here the aim is to keep the anther wall in one piece so that it can then be removed. Application of a coverslip followed by gentle squashing will make an ideal preparation.

4 The amount of squashing can only be learned through trial and error, but it is generally less than commonly used for mitosis. Little or no squashing is needed for studying general aspects of meiosis, where it is best to keep the original pollen mother cell wall intact. But for detailed study of chiasmata, orientation, chromatids, etc. it is necessary to apply firmish pressure to burst the cell wall and flatten the chromosomes.

5 Scan the preparation under the low power (x10) objective of the microscope. Note the flattened anther wall, some cells of which will be undergoing mitosis (a fact students need to be warned of before they spend unnecessary time studying mitosis again!). Surrounding the anther wall will be hundreds of pollen mother cells, many of which will be at the same stage of meiosis, and some non-dividing cells of the tapetum (Fig. 2.3d). Different anthers usually will be at different stages of meiosis, so that all the anthers on a slide should be checked. If the cells are too old or too young make another preparation, using a larger or smaller flower.

6 Remove any large air bubbles by applying 45% acetic acid (or propionic orcein) to the edge of the coverslip. Blot off excess liquid. Apply rubber solution or nail varnish to seal temporary preparations. Permanent preparation can be made as described for root tip cells (see Appendix).

General Procedures for Making Slides to Study Meiosis in Insect Material

Testes of locusts (or grasshoppers or crickets) provide very good material for studying meiosis in animals. Such material is not ideal to use when introducing meiosis to students, because the absence of a rigid cell wall means that cells and chromosomes are easily disrupted during slide preparation, making interpretation of stages and events of meiosis difficult. Also, the chromosomes of many insects are highly contracted at metaphase I, making their detailed study difficult. However, insect material is very good, some quite superb, for studying chiasmata at diplotene; and no study of meiosis would be complete without their inclusion. Locusts such as *Locusta migratoria* or *Schistocerca gregaria* are the best insects to use since they are easily bred in captivity and thus available at all times of the year. Alternatively, animals can be obtained from a biological supply house or, if really necessary, can be collected from natural populations during the summer when they will be sexually mature. Collect young adults or last-instar males and treat them as follows:

1 Anaesthetize the insect in a container with an absorbent pad soaked with chloroform or ether.

2 Hold the anaesthetized animal between thumb and forefinger, head inwards, and cut with scissors along the dorsal midline of the abdomen. With side pressure on the abdomen the testes, in some species, will pop out freely, or if not will be found on either side of the gut about half-way along the abdomen. Remove the testes with tweezers and place them in freshly prepared acetic alcohol. Each testis will consist of many banana-shaped follicles and, in some species, a quite extensive fat body. Any fat should be removed by combing the follicles lengthwise with a probe while holding the follicles in a bunch with fine tweezers in fixative in a petri dish. Transfer the cleaned follicles to a fresh lot of fixative and allow the material to stand for about 1 h. Material may be stored for long periods (years) in the fixative or in 70% alcohol in a refrigerator.

3 To make slides, remove a few follicles from the testis, drain them onto blotter and place them on a slide. Add a drop of propionic orcein (or aceto-orcein) and macerate the follicles with a blunt instrument (aluminium or glass rod). Discard any remaining large pieces of material. Apply a coverslip and allow the material to take up the stain for a few minutes, which may be enhanced by gentle heating of the slide over a spirit lamp or low bunsen flame. Material may also be stained with Snow's alcoholic carmine *before* making slides (as described for flower material), a technique that gives especially good results.

4 Squash the material between folded blotting paper. Temporary preparations sealed with rubber solution or nail varnish will last a number of days. Slides can also be made permanent (see Appendix).

5 Scan slides systematically. Amongst a variety of cell types, including maturing sperm, you will find spermatocytes undergoing meiosis, often in groups of ten or more cells all more-or-less at the same stage of division (Fig. 2.4). You may also find spermatogonial cells undergoing mitosis!

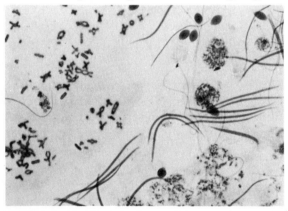

Fig. 2.4 Low magnification photograph from a testis squash preparation of *Locusta migratoria* showing spermatocytes in meiosis (left of centre) and maturing sperm.

EXERCISES

Meiosis is very often misunderstood and misrepresented, even in textbooks on genetics, though there is really no need for this. The exercises given below are designed, therefore, to give students a clear understanding of meiosis and its significance. We emphasize the importance of first having a good understanding of mitosis; and of the need for a good period of time to be given between studying mitosis and meiosis, to allow things to gel. Adequate time should then be given to allow unhurried study of meiosis.

1 Preparing and Studying Slides of Meiosis

General approach

Making slides of meiosis introduces students to where and when meiosis occurs in an organism; and then provides them with material with which to study the events of meiosis. Ideally, students should prepare and study slides of meiosis in a plant, such as *Allium triquetrum*, and in an animal, such as *Locusta migratoria*. To do this, follow the procedures given on pp. 33–34.

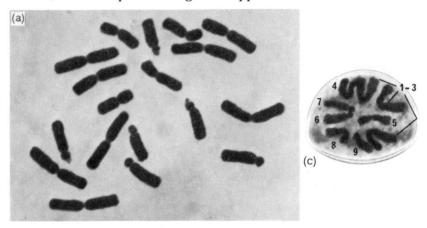

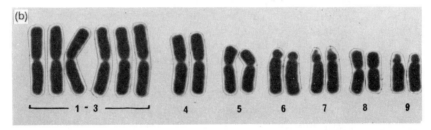

Fig. 2.5 The chromosomes of *Allium triquetrum* (2n = 2x = 18). (a) From a *diploid* root-tip cell in mitosis, treated with colchicine and stained with Feulgen's dye. (b) The same chromosomes as in (a) but after they have been cut out and aligned as homologous pairs in decreasing order of size. The centromeres are evident as constrictions across the chromosomes. Chromatids are not separately visible. The chromosomes fall into three main groups, namely (i) large, metacentric; pairs 1–4 (ii) small, sub-metacentric; pairs 5 and 8 (iii) small, acrocentric; pairs 6, 7 and 9. Within each of these groups certain chromosomes can be distinguished according to relative size, centromere position and nucleolus organizing regions (at the small terminal portions of the short arms of chromosomes 7 and 9). (c) The *haploid* chromosome set in a pollen grain undergoing its first maturation mitosis. Notice that there are now only nine chromosomes present, and only one of each homologous pair.

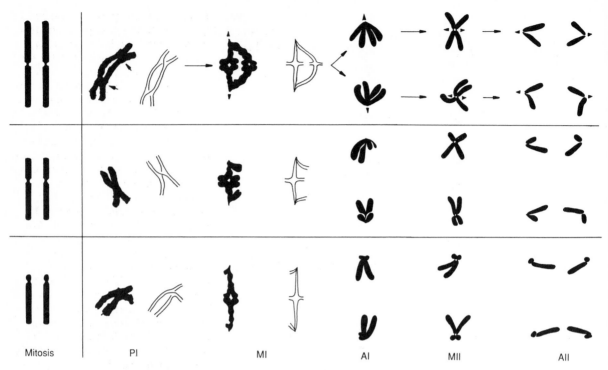

Mitosis   PI   MI   AI   MII   AII

Fig. 2.6 Drawings and line interpretations of the three main chromosome types in *Allium triquetrum*, as they appear in colchicine treated mitosis (see also Fig. 2.5) and in key stages of meiosis. Centromeres are not visible in prophase I but their approximate positions can be judged from their known location in mitotic chromosomes and in metaphase I bivalents. At later stages of meiosis centromeres are indicated for one chromosome type by single, tailless arrows, pointing towards the pole to which each is connected. Chiasmata at prophase I are indicated by double, tailless arrows. The second division (AII) is at right angles to the first division. P = prophase, M = metaphase, A = anaphase.

Before studying their slides students should be presented with a summary of the main events of meiosis, such as that given on p. 28. It will also be found useful to introduce students to the main types of chromosomes found in *mitotic* cells of the species being used, and then to suggest they try to find and study these different chromosome types at the various stages of meiosis. To this end, Fig. 2.5 is given here to indicate the three main types of chromosomes of *Allium triquetrum* as found in mitotic cells, and Fig. 2.6 is given to help in finding the corresponding chromosomes in meiosis.

Figs 2.7 and 2.8 illustrate meiosis in *Allium triquetrum*, as seen in slides prepared according to the technique given above; and Fig. 2.9 illustrates equivalent stages of meiosis in testis material of an insect.

In looking at their slides of meiosis, students should aim to study *events* not simply to try and identify the stages of meiosis. The main events to consider are chromosome contraction, orientation and separation; the origin and behaviour of bivalents, chiasmata and half-bivalents; and the first versus second divisions of meiosis. Special attention should be given to diplotene cells, particularly those of insect material, where chiasma structure and number should be studied using the

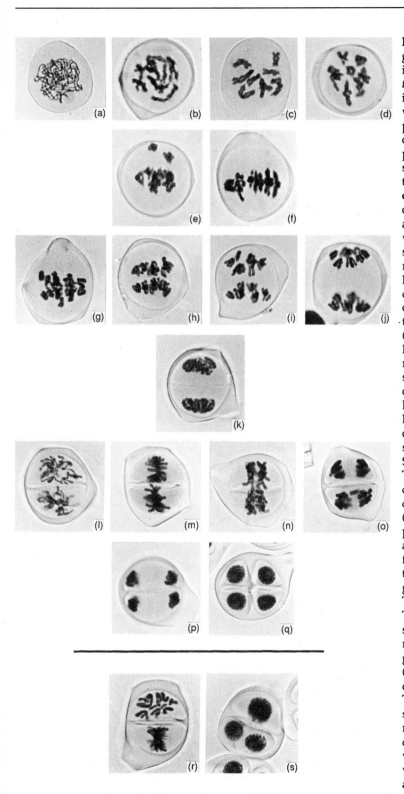

Fig. 2.7 Low magnification photographs of successive stages of meiosis in pollen mother cells of *Allium triquetrum,* fixed in acetic alcohol and stained in Snow's alcoholic carmine. The cells were only lightly squashed during slide preparation so as to preserve intact the original cell wall, which is very thick in places. The nuclear envelope and spindle fibres are not visible. The centromeres are not visible as such, but are easily located by the position and shape of the chromosomes at metaphase and anaphase. (a – d) Prophase I. In (a) which is the first easily recognizable stage of meiosis in this material, chromosome replication, synapsis of homologous chromosomes and chiasma formation have already occurred (*cf.* Fig. 2.2a–c). Contraction of the chromosomes has occurred between (a) and (c) and the first clear evidence of homologous chromosomes and chiasmata, though not chromatids, can be seen in (c). Two of nine bivalents partly overlap each other (lower centre). (e, f) Prometaphase I (e) in which two bivalents have yet to complete their orientation; and metaphase I (f). The spindle poles are N–S to the page. (g – j) Successive stages of anaphase I. (k) Telophase I/interphase (end of the first division). A new cell wall can be seen dividing the original cytoplasm in two. (l) Prophase II. (m) Metaphase II. The poles of the second division spindles are E–W (at right angles to that of the first division). (n, o) Anaphase II. Note the synchronous behaviour of the two groups of chromosomes. (p, q) Telophase II and the end of meiosis. The tetrad of young pollen grains are still contained within the original pollen mother cell wall. For liberated pollen grains, see Fig. 2.3e. (r, s) As for (m) and (q) except that the second division spindles are at right angles to each other. Thus in the lower cell of (r) the chromosomes would have divided E–W (as in m), but in the upper cell would have divided above and below the page, as it were. The consequence is shown in (s) where in the upper cell pair two nuclei are superimposed on each other and therefore appear as one.

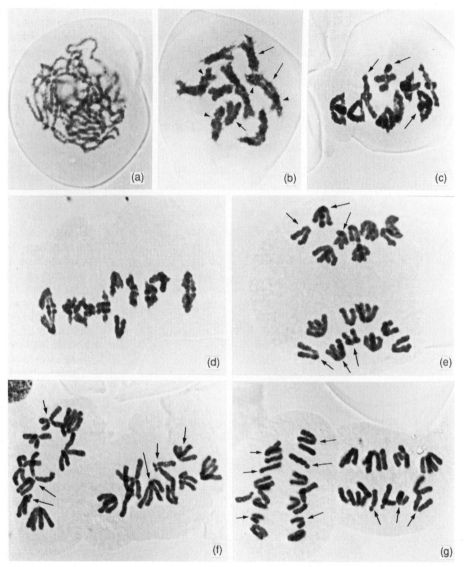

Fig. 2.8   Key stages of meiosis in *Allium triquetrum* shown at high magnification. During slide preparation the coverslip was firmly pressed to break the original cell wall and flatten and spread the chromosomes. The spindle poles are N–S in (c –e) and at right angles to each other in (f) and (g) The long arrows identify examples of the three main types of chromosomes (bivalents) found in *Allium triquetrum*, as illustrated and interpreted diagrammatically in Fig. 2.6. (a) Pachytene (prophase I). Each visible chromosome 'body' is in fact a bivalent, in that replication and synapsis has already taken place. Individual chromatids and homologous chromosomes cannot be seen. (b) Diplotene (prophase I). Centromeres are not recognizable at this stage. Thus chromosomes 5 and 8 are not distinguishable from 6, 7 and 9 (see Fig. 2.6), these chromosomes together making up the group of small bivalents seen at diplotene. Each bivalent has one or more chiasmata (tailless arrows). Chromatids are not distinguishable. (c) Metaphase I. The nine bivalents have been squashed a little out of alignment. Three bivalent types (Fig. 2.6) are now dis-

tinguishable. (d) Early anaphase I. One bivalent (centre) has completely separated into its two half-bivalents; another two bivalents (right and left sides) have nearly done so. (e) Anaphase I. Three half-bivalent types are distinguishable. (f) Metaphase II. Note here, as in (e), that the chromatids are well separated from each other, except at their centromeres. (g) Anaphase II. Nine chromatids (chromosomes) are present in each group. Centromeres here (and in f) are identifiable as weakly stained bands or constrictions across the chromosomes.

highest magnification of the microscope. Students should be encouraged to draw actual cells as they see them down the microscope; and then alongside each drawing to construct an interpretative diagram.

For some situations there may be insufficient time for students to study all the main events of meiosis in their own slides. Demonstration slides should be made available for these situations. A closed circuit television with a video-camera mounted on a microscope is very useful for pointing out to a large class key issues from good demonstration slides. Another good approach is to use photographs of individual cells with well spread chromosomes for detailed study by students.

Understanding meiosis
In studying meiosis from slides or photographs the following issues should be considered.

Meiosis as a whole
The first interesting point to notice from Fig. 2.7 is that the entire process of meiosis, i.e. *both* of its divisions, takes place within the original pollen mother cell wall. This is typical of meiosis in plant material. The original cell wall is easily ruptured when cells are flattened under thumb pressure during slide preparation; and in life it eventually ruptures of its own accord to release the maturing pollen cells.

Since there is no cell wall in animals the cells produced by the first division of meiosis are not contained within a common wall and thus often separate widely from each other, especially during slide preparation. Only in a few cases is it possible to find pairs of cells that certainly are sisters derived from the first division of meiosis (Fig. 2.9). Thus it is sometimes necessary to look closely at the number and nature of the chromosomes to distinguish cells at (say) anaphase I and anaphase II.

Chromosome replication and synapsis
By the time chromosomes are individually very distinct, such as in Figs 2.7a and 2.8a, homologous chromosomes have already replicated, synapsed and formed chiasmata. Despite this, only apparently *single* chromosomes can be seen at this stage. We are confident that these events have indeed occurred because of the way the chromosomes appear at later stages of meiosis. Thus by middle prophase I the haploid number of bivalents are present, each with one or more chiasmata (Fig. 2.8b); and in favourable material each bivalent will be seen to be composed of four chromatids (see especially Fig. 2.9h). These four chromatids are

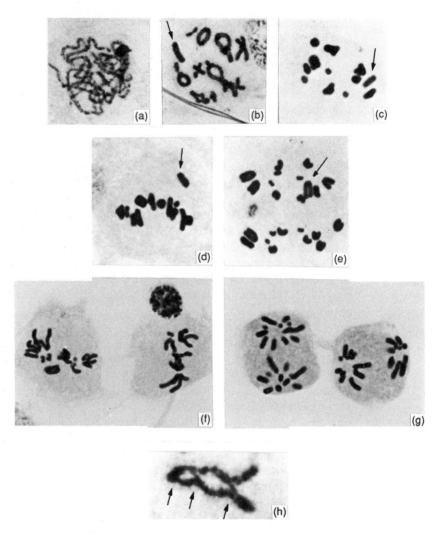

Fig. 2.9 Key stages of meiosis in *Locusta migratoria* . (a) Pachytene (b) Diplotene (c) Diakinesis (d) Metaphase I (e) Mid-anaphase I (f) Metaphase II (g) Anaphase II. The arrows locate the univalent (unpaired) X chromosome which, being without a partner, does not orient in the equator at metaphase I, but instead is located close to one pole (d). The spindle poles for the second division are at right angles to each other (f,g). (h) High magnification photograph of one diplotene bivalent showing three chiasmata (arrows). The lower chromatid at each of the two right hand chiasmata is out of focus.

the products of replication of a pair of homologous chromosomes, which have synapsed to produce the foursome of chromatids in a bivalent.

As for mitosis, chromosome replication for meiosis produces from a single chromosome a pair of sister chromatids that are genetically identical, each member of a pair carrying the same set of gene loci and alleles as the other (see Chapter 1, Fig. 1.5).

In bivalents of prophase I, therefore, we can see the *results* of chromosome replication. We can also see the results of synapsis of homologous chromosomes, though this is because by now homologous chromosomes are starting to separate from each other (Fig. 2.8b).

As yet we do not clearly understand how homologous chromosomes recognize each other and move together to produce a bivalent, partly because the events take place long before the chromosomes become individually distinct. We would like to know more about the process because synapsis sets the stage for the subsequent events of meiosis.

Chiasmata

In Fig. 2.8b the chiasmata appear simply as regions where homologous chromosomes of a bivalent remain held together. But in certain organisms, especially insects, where individual chromatids of a bivalent can be easily seen, a chiasma is then revealed as a cross-shaped arrangement of two of the four chromatids (Fig. 2.9h). In these bivalents it is possible to appreciate the fact that homologous chromosomes continue to be linked together at this stage of meiosis because of cohesion between sister chromatids on either side of the chiasma. The structure and significance of chiasmata form the basis of Exercise 2 below.

Bivalents at metaphase I

Cells showing chromosomes in the *process* of attachment to the spindle and orientation in the equator, such as that of Fig. 2.7e, are not easy to find, because this movement phase of meiosis is relatively short in duration. But metaphase I cells such as Fig. 2.7f are easy to find and deserve special attention, especially when flattened out such as in Fig. 2.8c. Here, different bivalents take on different appearances, depending on chromosome length, centromere position and location of chiasmata. In all cases the *chiasmata* lie along the equator, *not* the centromeres. The centromeres are situated some distance from the equator, pointing to opposite poles of the spindle. This is because in the first division of meiosis sister chromatid centromeres *behave as a unit*, rather than as independent entities (cf. for mitosis). This vital feature of meiosis means that in a bivalent, homologous chromosomes, not individual chromatids, become attached to opposite poles of the spindle at metaphase I. The centromere units in metaphase I then move as close to the poles as is permitted by the nearest chiasma, while the cohesion between sister chromatids at this nearest chiasma prevents the centromeres from moving immediately all the way to the poles. These very special features of meiosis make metaphase I unmistakably distinct from metaphase of mitosis, and require special attention when teaching meiosis in a laboratory.

The spindle cannot be seen in cells prepared according to the procedures given. Students might be encouraged, however, to include spindle fibres in their interpretative drawings, to be sure that they do indeed understand the manner in which bivalents are oriented at metaphase I.

Separation of half-bivalents in anaphase I

Fig. 2.7g–j shows the gradual separation of half-bivalents to opposite poles of the spindle during anaphase I, which students will find especially attractive to study, particularly in cells in which the original cell wall has not been broken. Early and later stages of this separation should be studied in flattened cells, such as those of Fig. 2.8d and e, respectively. It is during these stages that the individual chromatids of each half-bivalent become especially easy to see. This is because anaphase I of meiosis is triggered (as in mitosis) by a loss of cohesion between sister chromatids, which therefore easily splay apart from each other, as shown in Fig. 2.8e. Half-bivalents at anaphase I then take on different appearances depending on centromere position and chromosome size. These should be studied carefully.

It is important to note that, in contrast to anaphase of mitosis, the loss of chromatid cohesion at the start of anaphase I occurs all along the chromatid arms but *not* at the centromeres, which therefore do not uncouple. This ensures that a bivalent separates during anaphase I into half-bivalents, not into individual chromatids.

The second division of meiosis

Fig. 2.7k shows that when anaphase I has been completed the chromosomes usually form into a pair of nuclei and the cytoplasm divides to form two complete cells. These two cells then proceed in a more-or-less synchronous manner through the second division of meiosis, which make for very attractive viewing (Fig. 2.7l–p). Some cells will be found in which the axes of the second division are at right angles rather than parallel to each other (Fig. 2.7r). Corresponding cells at the end of meiosis (Fig. 2.7s) will puzzle students until they focus carefully through to find the 'missing' member.

Students will find that the second division of meiosis looks a little like mitosis, but in detail they are quite distinct. There is, of course, no chromosome replication in the interphase immediately prior to meiosis II. Chromatids for this division are already present, and can easily be seen in flattened metaphase II cells because sister chromatids are often splayed apart, as a result of the loss of their cohesion at the initiation of anaphase I (Fig. 2.8f).

At metaphase II, chromatid centromeres can be seen as independent though connected units (Fig. 2.8f), attached to opposite poles of the spindle. At the start of anaphase II chromatids lose their final cohesiveness (round their centromeres), allowing them to uncouple and move to the poles (Fig. 2.7n, o and Fig. 2.8g).

It should also be pointed out to students that since chromosomes replicate prior to the start of meiosis there has to be *two* divisions to meiosis, because the products of meiosis are required to contain the haploid number of *unreplicated* chromosomes. The second division is as crucial as the first in achieving this, a point that is often not grasped. Some organisms that reproduce parthenogenetically illustrate the point well, because for them meiosis is halted after the first division. The resulting gametes (eggs) develop without fertilization to produce adults having the *same* number of chromosomes as its parent had, not half this number.

2 Understanding the Nature and Significance of Chiasmata

Understanding the nature of chiasmata is perhaps the most difficult aspect of meiosis. Prophase I bivalents, particularly of insect spermatocytes (Fig. 2.10), therefore, deserve special study. Fig. 2.11 will help in this respect, as will the use of poppet bead or wire models (Fig. 2.12).

We now know that each chiasma results from *breakage* at corresponding points

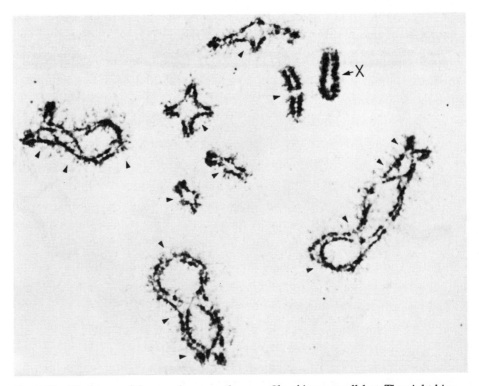

Fig. 2.10 Diplotene of the meadow grasshopper *Chorthippus parallelus* . The eight bivalents present have a total of 15 chiasmata (tailless arrows). The X chromosome (X) is a
univalent, and thus has no chiasma. The single chiasma of each of the five smaller bivalents is in 'cross-road' form (see Fig. 2.11a). Note the relationship between chiasma
number and chromosome length, except that even the smallest bivalent has one chiasma.
Also notice the more-or-less even distribution of chiasmata within the three large bivalents; and the different chromatids involved in successive chiasmata. Photograph courtesy
B. John.

of *non-sister* chromatids, followed by cross-wise *rejoining* of broken ends to form
new combinations of whole chromatids, as illustrated in Fig. 2.11a, 1–3. This
breakage and rejoining corresponds to the genetic phenomenon of crossing over.
A chiasma is the cross-shaped arrangement of the chromatids resulting from
crossing over. When we look at a chiasma, therefore, we are looking at a region at
which a rearrangement of genetic information in maternal and paternal chromosomes has taken place!

Students should ensure that when they draw a chiasma they *keep sister chromatids adhering to each other*, at every point. They can then produce a chiasma only by
breakage and rejoining. Representing a chiasma in the way shown in Fig. 2.11b is
incorrect. In teaching students the origin of a chiasma it is helpful to use poppet
bead (or wire) models, with different colours for homologous chromosomes and
identical colours for sister chromatids (Fig. 2.12).

For large chromosomes, successive chiasmata are spaced out along the bivalent, with each chiasma appearing in 'side' view (Fig. 2.10). However, especially
when there is only a single chiasma in a bivalent the paired chromatids usually

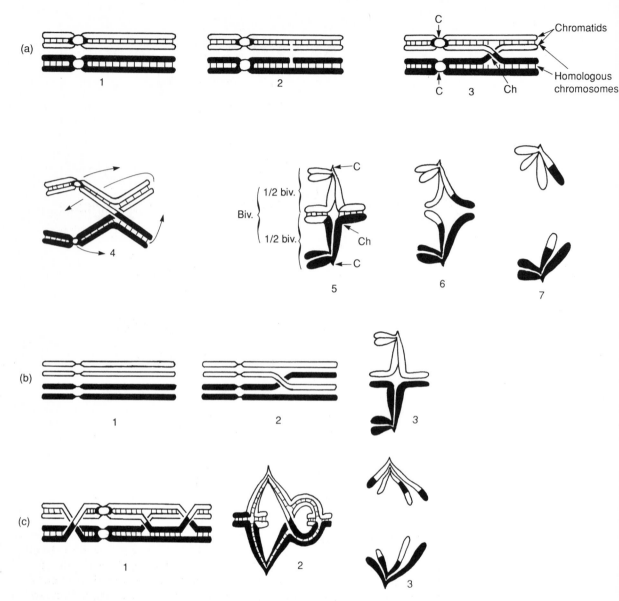

Fig. 2.11 Diagrams illustrating the origin and behaviour of a chiasma. (a) Illustration of how a single chiasma is formed. 1, after chromatid formation and pairing of homologous chromosomes, but prior to chiasma formation. 2, breakage of non-sister chromatids in corresponding positions. 3, rejoining of broken ends to produce new, whole chromatids. The chiasma so formed appears in 'side' view. Often, however, the chromosome arms rotate around the chiasma to give it an opened-out, 'cross-road' appearance, as shown in 4, 5. The chiasma is 'lost' (resolved) during anaphase I after the cohesion between sister chromatids (vertical lines) lapses and the centromeres move polewards (6 and 7). The exchange of chromosome material that has taken place is clear in 7. c = centromeres; c/tids = chromatids; hom.c/somes = homologous chromosomes; ch. = chiasma; biv. = bivalent. (b) Illustration of how *not* to draw a chiasma! Note that in 2, non-sister chromatids have merely exchanged pairing partners; there has been no exchange of material between homologous chromosomes (chromatids), nor can there be in anaphase. (c) A bivalent in which all four chromatids have been involved in the formation of three chiasmata; and the corresponding bivalent at metaphase I (1) and anaphase I (2 and 3).

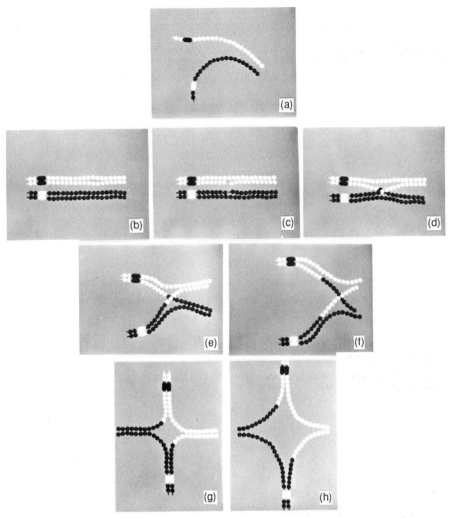

Fig. 2.12 Poppet bead models illustrating their use in understanding the origin and behaviour of a chiasma. One pair of homologous, acrocentric chromosomes is shown in black and white. (a) Prior to chromosome replication and prior to the start of meiosis. (b) After replication and pairing, but before chiasma formation. (c) Start of chiasma formation, i.e. breakage of non-sister chromatids at corresponding positions. (d) Rejoining to form new combinations of whole chromatids, which therefore lie one over the other at the resulting chiasma. (e) Metaphase I. Centromere pairs are shown as if connected to opposite poles of the spindle (N-S to the page). Homologous chromosomes (pairs of sister chromatids) have separated up to the point of the chiasma, where cohesion between sister chromatids prevents further separation until the start of anaphase. Strictly, sister chromatid centromeres should be shown one on top of the other, connected as a unit to the one pole. (f) Start of anaphase I. Cohesion between sister chromatids has lapsed, allowing the centromeres to separate freely towards opposite poles. Note that the chiasma is shunted towards the ends of the chromosomes during separation of the centromeres. The chiasma will eventually be 'lost' after further movement of the centromeres to the poles. (g & h) Same as (e) and (f) but after rotation and thus opening out of the chiasma. Strictly, sister chromatid centromeres should be shown rotated so as to face a pole.

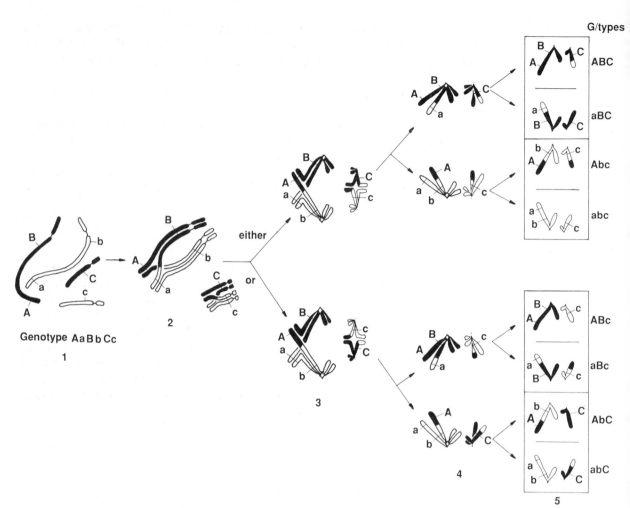

G/types

ABC
aBC
Abc
abc

ABc
aBc
AbC
abC

Genotype AaBbCc

1

2

either

or

3

4

5

Fig. 2.13 Simplified diagram of meiosis illustrating how chromosome behaviour underlies the three main laws of genetics. (1) chromosomes (only) prior to chromatid formation and prior to synapsis; (2) bivalents at prophase I (diplotene); (3) metaphase I; (4) anaphase I; (5) anaphase II and genotypes of the cells produced. Two pairs of acrocentric chromosomes are illustrated. The assigned genotype (**AaBbCc**) includes two gene loci that are linked and located in the long arms of the large bivalent (see part 1). At diplotene a chiasma is located *between* the two gene loci **A/a** and **B/b**; another chiasma is located on the non-centromeric side of the locus **C/c** in the small bivalent. At metaphase I, two possible orientations of the two bivalent are shown as 'either/or'. The corresponding segregations at anaphase I and anaphase II are then shown. Important points are: (i) Separation (segregation) of alleles into different cells takes place during anaphase I when crossing over (chiasma formation) does not take place between a gene locus and its centromere (for **B/b** and **C/c**); but is not completed until anaphase II if crossing over does occur in this region (for **A/a**). (ii) Random orientation of different bivalents at metaphase I accounts for the independent separation (segregation) and recombination of genes located in different (non-homologous) chromosomes. Independent segregation can also take place through different relative orientations of the chromosomes (half-bivalents) at metaphase II; for simplicity this is not illustrated, though it can be verified by including a fourth gene (**D/d**, say) distal to the chiasma in the small bivalent. (iii) Recombination of genes carried in the same chromosome (**A/a** and **B/b**) takes place when a chiasma forms between them.

rotate around the chiasma as if to release tension. The chiasma therefore 'opens out' to give it a 'cross-road' appearance (Figs 2.10 and 2.11a, 4–5), which is often wrongly interpreted as two chiasmata. The use of models conveys the correct situation very readily (Fig. 2.12).

Fig. 2.11c shows how *all four* chromatids can participate in the formation of successive chiasmata.

During early anaphase I sister chromatids peel apart from each other at the chiasmata (Fig. 2.11a, c), which are thus 'lost'. Poppet bead models again help a great deal in understanding this event (Fig. 2.12).

Chiasma formation means that the sister chromatids that separate during the second division of meiosis are genetically *dissimilar*, an important point that should be considered at some stage in this exercise.

3 The Results and Genetic Implications of Meiosis

Meiosis should be studied in a laboratory not in isolation but rather in relation to its outcome and its genetic implications. This can best be achieved as a supplementary exercise, by reference to photographs and diagrams. Thus meiosis in a diploid organism produces cells in which only *one of each type* of chromosome is present, whereas prior to meiosis, cells were diploid with two of each chromosome type (Fig. 2.5). The products of meiosis are also *genetically dissimilar*, because genetic information from the original parents of the organism concerned has been segregated and recombined to produce novel combinations. This segregation and recombination occurs because of the behaviour of chromosomes during meiosis (Fig. 2.13).

REFERENCES

Dyer A.F. (1979). *Investigating Chromosomes*. London, Edward Arnold.
John B. and Lewis K.R. (1984). *The Meiotic Mechanism*. Carolina Biology Reader, ed. J.J. Head. Edinburgh and London, Carolina Biological Supply Company.

Chapter 3 DROSOPHILA GENETICS

The exercises described in this chapter are suitable for all levels of study, from senior school to university undergraduate. Some choice, however, should be practised so as to use the more simple exercises for lower levels of study.

Drosophila are commonly occurring wild flies and are harmless to humans. The medium used for culturing the flies is also free of hazards. The chloroform or ether used to anaesthetize flies, however, presents both a fire and health hazard. Thus ether should be used in the total absence of any naked flame; and windows and doors of the laboratory should be opened to avoid build up of fumes. Bulk supply of ether should be kept in and dispensed from a fume cupboard (hood), ideally in a preparation room separate from the laboratory itself.

Occasionally culture bottles and tubes and other glasswear can become cracked or broken during normal use. Such defective glasswear is a hazard and should be properly disposed of immediately.

INTRODUCTION AND BACKGROUND

The fruit fly, *Drosophila melanogaster*, is an excellent organism for experiments on heredity. It was first brought into use for this purpose by T.H. Morgan in 1910 and has been growing in popularity and importance ever since.

Drosophila is an insect. It is a diploid organism with the low chromosome number of 2n = 8. It has a short generation time, produces large families of offspring, is easy to handle and has many mutant characters with which to work. Three other important features are its XX/XY chromosome system of sex determination, its giant chromosomes in the salivary glands of its larvae, and the fact that crossing over at meiosis does not occur in the male.

Drosophila eggs are 0.5 mm in length and white in colour. The ones that have been fertilized hatch into larvae 24 h after laying. Females that have not been fertilized will also lay eggs, but these will not develop. The larvae feed voraciously while burrowing through the food on which they are being raised, and grow to a size of 4.5 mm. When they are ready to pupate they crawl out of the culture medium and find a dry place to attach themselves to. The pupa develops within the skin of the larva and towards the end of pupation the form of the adult can be seen through the larval skin. When the adult emerges it is elongate, pale in colour and has crumpled wings (a condition that some students are at first keen to regard as representing a new mutant!). It takes about an hour for the wings to expand fully and to dry out and for normal body shape to be achieved. After 12 h the pigmentation on the body of the fly is fully developed. Females remain virgin for only about 6 h after emergence, and start to lay eggs when they are 2 days old.

PROCEDURES

Culture Medium

In natural populations fruit flies feed off soft fruits, with a preference for those that are overripe and fermenting. In the laboratory they live well on an artificial medium which contains agar and which therefore remains solid enough to allow flies to be tipped out of the culture vessel for handling. Flies can be grown in small, flat-bottomed specimen tubes (75 x 25 mm), or in 1/3 or 1/4 pint-sized milk bottles. Instant Drosophila medium can be purchased from some biological supply houses and reconstituted simply by adding water. On the other hand, it is not difficult to make up an appropriate food for use as required. The medium given below is one we have used successfully for many years. The ingredients can be easily and cheaply obtained and are as follows:

Cornmeal (maize meal)	60 g
Soya meal	6 g
Agar	5 g
Sugar syrup	60 ml
Dried yeast	3 g
Water	500 ml
Nipagin	pinch
Propionic acid	2.5 ml

The above makes enough food for 10–12 1/3 or 1/4 pint-sized milk bottles; or about 40 tubes. Sugar syrup is made by adding 1 kg of sugar to 500 ml of cold water and boiling for a while. Nipagin is *methyl-p-hydroxybenzoate* . With propionic acid it inhibits the growth of fungi, bacteria and mites. Mites can be particularly troublesome once they become established in a culture.

The procedure for making up the culture vessels is as follows:

1 Have the bottles or tubes ready sterilized by heating in an oven at 160°C for 1 h, or by autoclaving at 10 p.s.i. for 10 min.
2 Mix the ingredients of the food in a saucepan and let it boil for 1–2 min while stirring. Add 2.5 ml of propionic acid and pour cleanly into the bottles (or tubes) to a depth of about 2–4 cm.
3 Allow the medium to cool and solidify and then insert a stopper made of cotton wool tied in muslin or a foam rubber bung. Allow the bottles to dry for 48 h.
4 Just before use seed the surface of the food with a few grains of powdered inactive yeast.

Facilities needed for keeping Drosophila are modest. The biggest demand is time, and for an incubator in which the cultures can be maintained at 25°C. If an incubator is not available, flies can be reared at room temperature in the laboratory, or in a light box. A good supply of 1/3 or 1/4 pint-sized milk bottles and specimen tubes is also needed. A low-power (x10 magnification) stereomicroscope or hand lens and a fine haired brush are desirable items of equipment. Items needed for student handling of flies are given in the following sections.

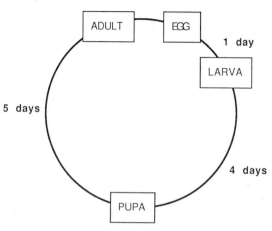

Fig. 3.1 The life cycle of *Drosophila melanogaster*, which is completed in 10 days at a temperature of 25°C.

Culture Conditions

The optimum temperature at which to keep fruit flies for practical work is 25°C. At this temperature the flies complete the four phases of their life cycle in about 10 days (Fig. 3.1). At lower temperatures the cycle is prolonged, and at higher ones it is completed more quickly. The maximum temperature at which cultures should be kept is 28°C because above this temperature males are sterile. The life cycle is prolonged in most mutant forms by 1 or 2 days.

Handling Flies

Etherizing

Flies are anaesthetized with diethyl ether in order to keep them passive whilst they are being examined or transferred from one culture vessel to another. The etherizing procedure is illustrated in Fig. 3.2. It may be carried out using purchased equipment, as shown in Fig. 3.2, or with home-made equipment. A simple etherizing device for class use is a conical flask fitted with a plastic filter funnel and, at the ready, a fitting cork with a small pad of cotton wool attached to the lower surface of the cork with a drawing pin. When etherizing flies first make sure that the flask is dry and free of dead flies from previous use. Tap the culture vessel on the side to force the flies away from the neck, and then quickly remove the bung and invert the vessel into the funnel. Keep all three parts of the system together and gently tap the base of the etherizer on the bench top, using a rubber pad or notebook to absorb the shock. The flies will fall through the funnel and into the etherizer (but avoid tapping too violently, otherwise the food will fall through as well !). Remove the culture vessel and re-bung. The flies are now trapped inside the etherizer. Add several drops of ether to the cotton wool pad on the cork, and after tapping the flies down from the neck of the etherizer quickly replace the funnel with the cork. Alternatively, add ether vapour via a soft plastic squeeze bottle, in the bottom of which is put a small quantity of ether and from the top of which a plastic nozzle projects. The nozzle is inserted into the container of flies and the bottle is squeezed to add ether vapour.

Fig. 3.2 Equipment used in handling Drosophila flies.

With either technique for adding ether the flies will fall to the bottom of the etherizer and stop moving after a minute or so. When fully asleep tip the flies out onto a white card or into a petri dish base lined with filter paper. Excessive ether will kill the flies (dead flies have their wings outstretched and legs drawn up). A small paint brush is desirable for moving the flies about. They can be kept under the anaesthetic for up to 30 min by re-etherizing. Re-etherizing is done with the petri dish lid which has a small square of absorbent paper taped on the inside. Add ether liquid to the paper and invert the dish lid over the flies to re-etherize them. Be especially wary of over-etherising them at this stage. Flies that are finished with can be disposed of in a morgue, which is simply a wide necked bottle containing dilute liquid detergent.

Sexing

In mature adult flies the main differences between the sexes are obvious, even to the naked eye (Fig. 3.3a). The female has a larger and more pointed abdomen than the male, particularly when she is distended with eggs. She also has several bands of dark markings on her abdomen, whereas the male has fewer bands and his abdomen is short and blunt with a much darker 'tail end'. The male is also characterized by the presence of a *sex comb*, consisting of about 10 dark bristles on the uppermost tarsal joint of the foreleg (inset of Fig. 3.3a). This sex comb is a diagnostic feature of the male and can be used to distinguish the sexes in the first 2 hours after hatching, when the form and pigmentation of the body are not fully

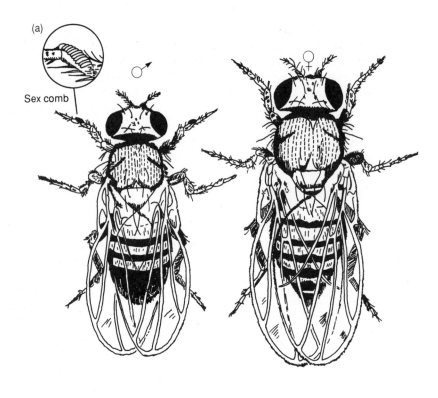

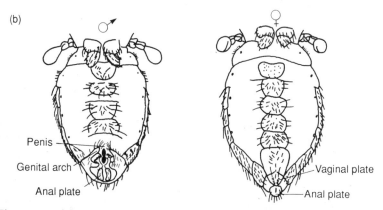

Fig. 3.3   (a) Male and female forms of adult *Drosophila melanogaster*. The inset shows the sex comb of the male at higher magnification. (b) Diagram showing differences in genitalia of male and female Drosophila.

developed. To see the sex comb it is necessary to use a stereomicroscope or a high-power hand lens. Caution is needed in relying on the *absence* of the sex comb for identifying females (you may simply not be seeing it !), especially if these are to be used to set up a cross. For absolute certainty students should distinguish the two sexes according to differences in their genitalia (Fig. 3.3b).

Collecting virgin females

To make breeding experiments with Drosophila it is necessary to collect females before they have mated, because females store sperm. To obtain virgin females, bottles of young cultures are emptied of *all* flies (check and re-check!) first thing in the morning. Later that same day the bottles are emptied again and the sexes separated. Females which have hatched during the day and which are not more than 5–6 hours old when separated from the males will be virgin. They can be collected into fresh tubes of food and kept for a few days in readiness for breeding experiments. It is best to share them around several tubes and then keep them unmated for a few days, in case any fertilized ones, or males, should slip through. These will be noticed when eggs in the tube develop into larvae.

Maintaining Stock Cultures

Strains of various mutants and of normal (wild type) flies may be purchased when required, or they can be maintained in the laboratory as stock cultures. When maintaining cultures each mutant strain is kept as a pure breeding line. Five to ten pairs have to be transferred to fresh bottles of food every 3 – 4 weeks. It is usual to keep at least two bottles of each strain, for safety reasons. In setting up transfers the new cultures should be checked for purity and each bottle clearly labelled to show which mutant it contains. To do a transfer, etherized flies are placed on a small paper 'boat' which is then slipped into a bottle or tube. The bottle or tube is laid on its side to prevent flies becoming stuck to the food. When the flies are active again the container can be turned upright.

For purifying contaminated cultures see the Appendix.

Setting up a Cross

The procedure here is essentially the same as that used for maintaining stock cultures, except that pairs of flies representing the parents of the cross to be set up are collected and placed together on the paper boat. For $1/4$ or $1/3$ pint-size milk bottles usually four to five pairs are used: when many more are put in the bottle the culture becomes overcrowded with larvae. Similarly, for small tubes two–three pairs of flies are used to establish a culture.

Label the culture vessel and seed the food with yeast *before* placing the flies into it; it is difficult to do it afterwards! Stand the culture vessel upright only when the flies have recovered from the ether. Incubate the culture at 25°C unless an experimental procedure states otherwise.

Symbols and Descriptions of Commonly Used Mutants

With Drosophila it is usual to denote the wild type (normal) allele of a gene at a particular locus by a + . A mutant allele is given a symbol which is the first one or first two letters of the word which describes the mutation. Thus *bw* is the symbol for the brown eye allele, *vg* for the vestigial wing allele and *w* for the white eye allele. The corresponding wild type alleles can all be given the symbol +, or they may be distinguished by writing *bw+*, *vg+* and *w+*, respectively. Recessive mutant alleles are denoted by lower case letters (such as *vg*), dominant ones by upper case letters (such as *B* for bar eye and *B+* for its wild type allele).

These symbols can also be used to denote the wild type or mutant *phenotype*. Thus + is used to indicate the wild type phenotype; *w* for the white eye phenotype; *bw* for the brown eye phenotype. Used carefully this system can save a lot of writing when describing the parents and progeny of a cross. The context makes it clear whether a given symbol refers to an allele or its phenotype. In some publications, symbols for phenotypes are written in plain text rather than in italics.

Symbols are also used to write a cross. Thus if we cross vestigial winged flies with wild type (normal, long winged) flies and wish to represent the *phenotype* of the parents we write:

$$vg \times + \quad or \quad + \times vg$$

These are *reciprocal crosses*, in the first of which the female is vestigial winged (and the male wild type) and in the second the female is wild type (and the male vestigial winged). By convention we always write the female parent first.

If in writing these crosses we wish to denote the *genotype* of the flies in question we write:

$$\frac{vg}{vg} \quad \times \quad \frac{vg^+}{vg^+} \quad and \quad \frac{vg^+}{vg^+} \quad \times \quad \frac{vg}{vg}$$

The double lines represent the homologous chromosomes in which the pair of alleles are located. We will mostly use this system in this chapter, since it is the best one to use for students beginning a study of genetics. This is because the behaviour of alleles is determined by the behaviour of their respective chromosomes. Alternatively, we can simplify the system either by removing one line and writing the alleles in sequence, such as $vg^+/vg^+ \times vg/vg$, or by removing both lines as in $vg^+vg^+ \times vg\ vg$. Premature use of the latter system can lead to a lot of confu-

Table 3.1 Characteristics of commonly used Drosophila mutants and their symbols.

Symbol	Name of mutant	Description of mutant
vg	vestigial	vestigial wing
w	white	white eye
st	scarlet	scarlet eye
se	sepia	sepia eye
m	miniature	miniature wings
bw	brown	brown eye
cn	cinnabar	cinnabar eye
B	Bar	Bar eye: eye reduced to a narrow slit
v	vermillion	vermillion eye
e	ebony	ebony body
f	forked	bristles gnarled and misshapen
y	yellow	yellow body
cu	curled	wings curled up at tip
b	black	black body

sion as to exactly what is being done when alleles are segregated into gametes and brought together again at fertilization. Segregation and recombination of alleles is much easier to understand if their behaviour is linked to the behaviour of their chromosomes.

The characteristics of some commonly used mutants are listed in Table 3.1.

EXERCISES
1 Monohybrid Cross: Segregation of Alleles

In a monohybrid cross we deal with a *single* pair of contrasting characters. In Drosophila a popular choice of mutant for use in such a cross is vestigial wing (*vg*). We cross vestigial winged flies with wild type (long winged) flies. Ideally *reciprocal crosses* should be set up. For these we need virgin females for both the vestigial line and the wild type line. The crosses to be set up are therefore: (i) vestigial wing (female) x long wing (male); *and* (ii) long wing (female) x vestigial wing (male).

The procedure to adopt is as follows:

1 Collect virgin females of both the mutant and wild type stocks.
2 Collect males. Label the culture vessel and seed the food with yeast. Set up matings by placing the appropriate male and female flies inside the culture vessel. Incubate.
3 After 3 days remove the parent flies. By this time a sufficient number of fertilized eggs will have been laid. It is necessary to remove the parents to avoid them mating with the F_1 when they hatch.
4 The F_1 flies can be examined 10 days or so after making the crosses. Note that all the F_1, both males and females, have the normal, long wing phenotype, irrespective of which way the cross was made. This result tells us that the normal wild type allele (vg^+) is dominant and that the genes are not sex linked (Exercise 3).
5 Place five (or two) pairs of F_1 flies into fresh containers of food. This can be done without selecting virgin females because the F_1 are all of one genotype and the purpose is simply to mate together the females and males. This is known as *intercrossing* the F_1 and is written F_1(female) x F_1(male), or simply F_1 x F_1.
6 After 3 days remove the F_1 flies and wait a further ten days or so for the F_2 to

Table 3.2 F_2 results from the Drosophila cross long wings (wild type) x vestigial wings; and results of a chi-squared test on them.

	Phenotypic class		
	Long wing	Vestigial wing	Total
Observed	390	114	504
Expected	378	126	504

χ^2 (3 : 1) = 1.52, degrees of freedom (*d.f.*) 1, P = 0.2, not significant.

emerge. The F_2 flies should be collected over a period of several days until all of the pupa cases are empty. This is because the mutant phenotypes usually take longer to hatch, due to their various developmental abnormalities.

7 Count the number of the two kinds of phenotypes in the F_2, males and females separately. They will approximate to a 3:1 ratio of normal (long wing) : vestigial. Actual results from a class experiment are shown in Table 3.2, along with those of a chi-squared (χ^2) test applied to them. A short account of the chi-squared statistical test for the analysis of genetic ratios is given in the Appendix.

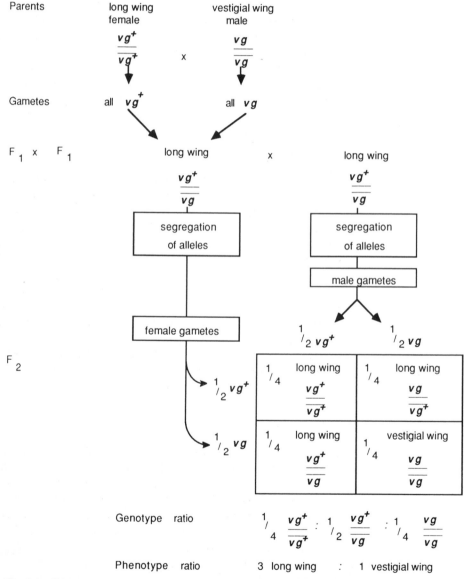

Fig. 3.4   Diagram explaining the F_1 and F_2 of the Drosophila cross long wing (wild type) x vestigial wing.

The results are explained in Fig. 3.4.

This exercise demonstrates the *principle of segregation*, i.e. the separation of the two alleles of the heterozygous F_1 into different gametes, resulting in the formation of two kinds of gametes in equal proportions, with each gamete containing only one allele (i.e. $\frac{1}{2}$ vg^+ and $\frac{1}{2}$ vg). This separation takes place because during meiosis of the F_1 the homologous chromosomes in which the two alleles are located separate into different cells, which become the gametes.

It is important to note that what we actually observe in this exercise is the phenotype of the parents and of the F_1 and F_2 flies. Our deductions about genotypes are based on the interpretation that the character is determined by a single gene with two alleles, and that one allele (vg^+) is dominant and the other (vg) recessive. The 3:1 ratio arises because the heterozygous F_1 flies produce two kinds of gametes in equal frequencies, and these gametes are equally viable and randomly combine into pairs during fertilization to give the F_2 progeny.

One of the major problems with using Drosophila in teaching genetics in a practical class is that of *differential viability* of flies of different genotypes. In reality, mutant genes seldom if ever have single effects on the overall phenotype of a fly. Rather they have multiple effects, many of which are detrimental to development and reduce the fitness of the mutant relative to the wild type phenotype. In the competitive environment of a culture bottle mutants often perform poorly and thus appear in less than their expected numbers. Especially if results from a large class are pooled and analysed together this differential viability can produce a statistically significant departure from the expected 3:1 ratio. This should not be ignored (and thus the exercise regarded as having failed) but rather should be taken as an opportunity to introduce students to the phenomena of pleiotropism (multiple effects of a gene) and differential viability, which are commonly observed.

Other mutant characters in Drosophila that give 'good' monohybrid ratios are ebony body (*e*) and sepia eyes (*se*). Crosses involving these characters can be treated in exactly the same way as for the vestigial wing cross. In a large class, variety can be introduced by allowing some students to set up the cross vestigial wing x wild type, others ebony body x wild type, still others sepia eye x wild type (and/or the reciprocal of each). At the end of the exercises results can be compared.

2 The Testcross

In the monohybrid cross described in Exercise 1 the genotypes of the gametes cannot be seen, and the principle of segregation is deduced indirectly from the ratio of phenotypes in the F_2. To confirm the genotypes of the F_1 we make a *testcross*, which mates the F_1 with the double recessive parent. Thus for our cross in Exercise 1, individual virgin females from the F_1 are mated with homozygous recessive males; or individual F_1 males with virgin homozygous recessive females. The heterozygous F_1 will produce two kinds of gametes, namely $\frac{1}{2}vg^+$ and $\frac{1}{2}vg$. The double recessive genotype will produce gametes of one kind, and since these are known to carry the recessive allele (vg) the phenotype of the testcross progeny will show up the genotype of the F_1. The explanation is given in

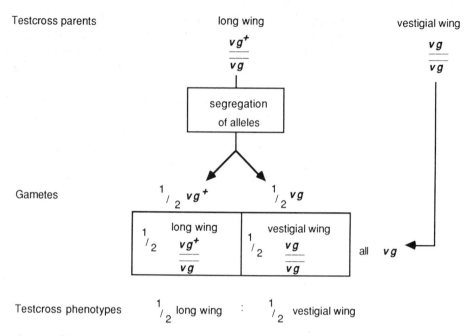

Fig. 3.5 Diagram explaining the results of a testcross of the F_1 from the cross long wing (wild type) x vestigial wing. The testcross confirms the genotype of the heterozygote.

Fig. 3.5. The 1:1 ratio of the testcross progeny should be confirmed using the χ^2 test.

3 Sex-linked Inheritance

Drosophila has an **XX/XY** chromosome system of sex determination. The male is **XY** and produces two types of sperm, carrying either the **X** or the **Y** chromosome. The female is **XX** and produces only **X** carrying eggs. It is important to realise that the **Y** chromosome is virtually devoid of genes and therefore males are described as *hemizygous*, in that they carry only one allele for each of the genes located in the **X** chromosome. This results in a pattern of inheritance for genes located in the **X** chromosome (X-linked genes) that is different from those in the autosomes.

Table 3.3 F_1 and F_2 results from reciprocal crosses involving white eye (*w*) and red eye (wild type) Drosophila.

	Cross number	
	(i)	*(ii)*
Parents:	white eye female x red eye male	red eye female x white eye male
F_1:	red eye females, white eye males	red eye male and females
F_2:	335 red eye, 315 white eye, male and females	703 red eye, 240 white eye; males only are white eyed

White eye (w) is a favourite character for demonstrating sex-linked inheritance. Other sex-linked traits in Drosophila are miniature wing (m) and yellow body (y). The procedure for setting up and analysing the results of the white eye cross is given below. Note that 'red eye' here refers to the dull red, wild type eye colour.

Reciprocal crosses are set up so that the mutant allele (w) is introduced from both the female and the male parent. Half the students in a class might set up the cross one way, and half the other way. These crosses are set up as described for the vestigial wing exercise. Typical results are shown in Table 3.3. Here we see that the phenotype of the F_1 depends on the sex of the individuals being looked at, from which we conclude that the w gene is *sex-linked*. This reciprocal differ-

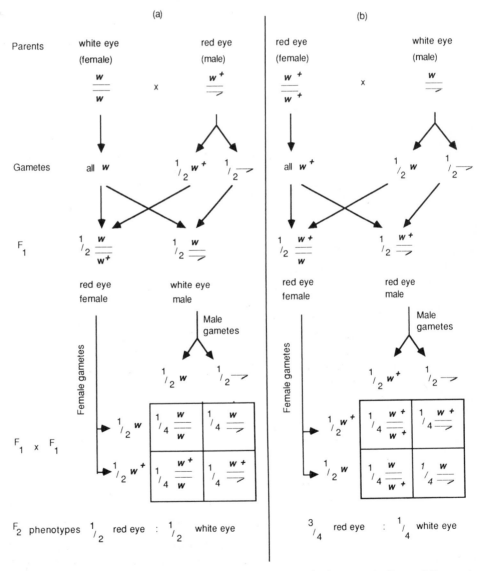

Fig. 3.6 Diagram explaining the inheritance of the sex-linked gene w in Drosophila.

ence is not found with autosomal genes (i.e. genes in chromosomes other than the sex chromosomes). Notice also the criss-cross pattern of inheritance in the first cross, namely white eyed mother → white eyed son, red eyed father → red eyed daughter.

In the F_2 of the first cross there is a 1 : 1 ratio between red eyed and white eyed flies, and males and females are alike in eye colour. But in the second cross a 3 : 1 ratio of red : white is found, and the only white eyed flies are male. The interpretation is given in Fig. 3.6, where the Y chromosome is represented by a 'half-arrow' symbol ⇀ and the recessive white eye allele as *w*.

4 The Dihybrid Cross: Independent Segregation

A dihybrid cross is one in which the parents differ in respect of *two* pairs of contrasting characters, the inheritance of which are therefore examined together. Such a cross is used to demonstrate the principle of *independent segregation* (= independent assortment) for two unlinked genes. The cross can be carried out using the autosomal mutants *vg* (in chromosome II) and *se* (sepia eye, in chromosome III), or other appropriate pairs of characters (see Table 3.1). The two mutants can be introduced into the cross from the *same* parent (such as by crossing vestigial wing, sepia eye flies with wild type); or one mutant can be introduced from each parent (vestigial wing, red eye x long wing, sepia eye). The results for these two crosses will be the same, because we are dealing with unlinked genes and the F_1 is simply heterozygous at both gene loci.

All the F_1 flies will be wild type since both mutations are recessive.

F_1 males and females are then mated to give the F_2 generation. The F_1 flies produce four kinds of gametes in equal frequencies by independent segregation ($\frac{1}{4}vg^+se^+$; $\frac{1}{4}vg^+se$; $\frac{1}{4}vg\ se^+$; $\frac{1}{4}vg\ se$). The gametes then combine at random in pairs to give four phenotypic classes in the F_2 in a ratio of 9 : 3 : 3 : 1. Actual results are shown in Table 3.4.

Table 3.4 F_2 results from the Drosophila cross vestigial wing, sepia eye x wild type; and results of a chi-squared test.

	Phenotype			
	+ +	+ se	vg +	vg se
Numbers:	1577	568	512	158
Ratio:	9 :	3 :	3 :	1

χ^2 (9 : 3 : 3 : 1) = 5.37, *d.f.* 3, $P = 0.15$, not significant.

Notice first that in Table 3.4 we have abbreviated the writing of the results of the cross by using a + for each of the wild type phenotypes (normal wing, normal eye). Thus in a + + fly, the first + indicates the wild type wing, the second + the wild type eye colour, and in the phenotype + *se* , for example, the + refers to the wild type wing.

The results of Table 3.4 are explained in Fig. 3.7.

In this exercise attention should be drawn to the process of *recombination* that takes place when two (or more) pairs of alleles are segregating independently

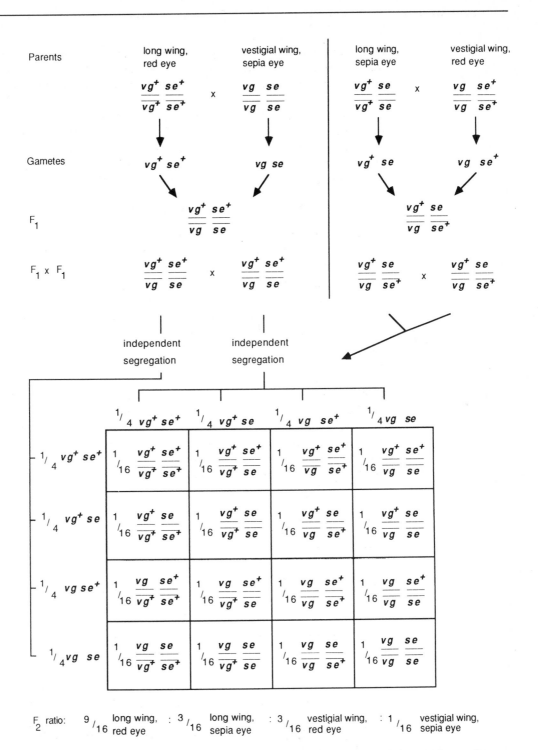

Fig. 3.7 Diagrammatic explanation of the F_1 and F_2 of the Drosophila cross long wing, red eye (wild type) x vestigial wing, sepia eye (left) and of the cross long wing, sepia eye x vestigial wing, red eye. The F_2 of these two crosses are the same because the two gene loci concerned are located in different chromosomes.

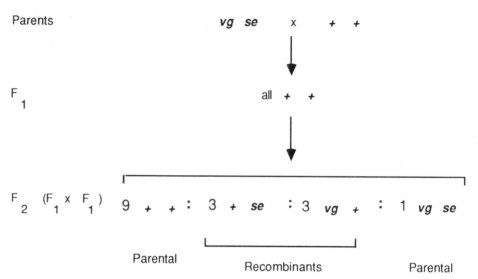

Fig. 3.8 Diagram identifying parental (= non-recombinant) and recombinant phenotypes found in the F_2 of the dihybrid cross vestigial wing, sepia eye x wild type. Note that for the cross *vg* + x +*se* the progeny types and their proportions would be the same as for *vg se* x ++ , but the parental and recombinant phenotypes would be reversed.

from a heterozygous F_1. The parents of the cross are of two kinds, but in the F_2 there are four different phenotypes. Two of these F_2 phenotypes are the same as the original parents (and are therefore called *parental* or *non-recombinant* types), and two have new combinations of characteristics by which they differ from the parents (and which we therefore call *recombinants*). Thus, for the *vg se* exercise above we can summarize the outcome in terms of recombination in the way shown in Fig. 3.8.

5 Linkage

Genes that are located in the same chromosome are said to be *linked*. Crosses involving such genes are used to demonstrate linkage, and to demonstrate the recombination of linked genes that takes place by *crossing over*. For this purpose we can use the autosomal gene vesitigial (*vg*) again, together with cinnabar (*cn*, bright red eye), which is also in chromosome II.

The following cross is made to obtain the heterozygous F_1 generation.

Parents: Wild type wing, wild type eye x vestigial wing, cinnabar eye, (symbolized + + x *vg cn*), or its reciprocal (*vg cn* x + +). Alternatively, the two mutant traits may be in both parents, one in each (*vg* + x + *cn* , or its reciprocal). Note that the F_2 for the latter two crosses will *not* be the same as for the first two (as explained later). The results given below are for the cross + + x *vg cn* or its reciprocal.

F_1:   All wild type for wing and eye colour (+ +), because both mutant alleles are recessive. Also, all the F_1 will be heterozygous for both pairs of alleles, as shown in Fig. 3.9.

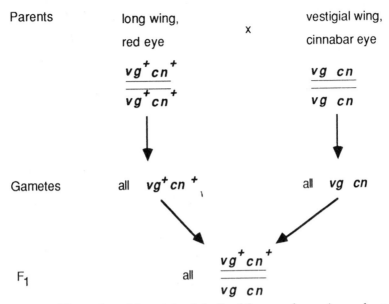

Fig. 3.9 Illustration of the origin of the F_1 of the cross long wing, red eye (wild type) x vestigial wing, cinnabar eye.

Now, in linkage experiments we usually do not proceed to the F_2 in the normal manner (F_1 x F_1); it is a lot more informative to examine linkage by making *testcrosses* with the homozygous mutant stock, i.e. (i) F_1 x *vg cn* and (ii) *vg cn* x F_1. In this way we can see directly what kinds of gametes are being produced by the heterozygous F_1, because their genotypes will be revealed in the phenotypes of the testcross progeny.

Thus, after collecting virgin females, testcrosses are made by using the heterozygous F_1 flies as both the female and male parents. Table 3.5 shows a typical set of class results for this exercise.

Table 3.5 Results of testcrosses of F_1 females and males derived from the Drosophila cross + + x *vg cn*.

		Testcross progeny (males and females alike)			
(i) F_1 (female) x vg cn (male)			(ii) vg cn (female) x F_1 (male)		
+ +	136	non-recombinant (parental)	+ +	144	non-recombinant (parental)
+ cn	15	} recombinants	+ cn	–	
vg +	14		vg +	–	
vg cn	119	non-recombinant (parental)	vg cn	141	non-recombinant (parental)
Total	284			285	

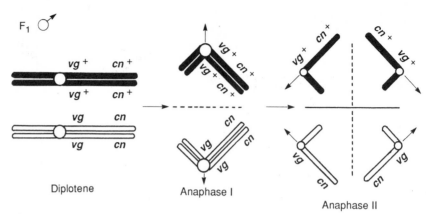

Fig. 3.10 Diagram showing segregation of chromosomes and thus the alleles they carry during meiosis of male Drosophila, in which there is no crossing over.

The first thing we notice in these results is the difference between the two crosses. When the male is the heterozygote (ii) we have *complete linkage* in that the parental combinations of characters (+ + and *vg cn*) are the only ones to be transmitted by the male to its progeny. There are no recombinants. The reason for this is that Drosophila (along with a few other organisms) is unusual in that there is *no crossing over at all* in the male. During meiosis in males, homologous chromosomes pair together side-by-side during the first prophase, and then separate in the normal way at anaphase but without any exchange of segments of chromatids, i.e. without chiasma formation having taken place (Fig. 3.10).

When the female is used as the F_1 heterozygote, however, we have four classes of progeny, two of which are parental types and two of which are recombinants. Linkage is evident because the four classes are not present in the 1 : 1 : 1 : 1 ratio that we would expect from independent segregation. Instead of 25% of each category the two parental types are far more frequent than the recombinant types, which amount to only 10% of the total progeny.

In order to confirm linkage we can look at the two pairs of alleles *separately* to show that there is no disturbance to their normal segregation, which might otherwise account for the result. For wild type wing : vestigial wing (+ : *vg*) the ratio is 151 (136+15) : 133 (119+14). For a 1:1 segregation ratio this gives a χ^2 value of 1.14 which, with *d.f.* = 1, is not significant. Similarly, for wild type eye : cinnabar eye (+ : *cn*) the ratio is 150 (136+14) : 134 (119+15). This gives a χ^2 value of 0.90, which is also not significant.

Where the female is used as the F_1 heterozygote, therefore, we have *partial linkage*, which is the kind of linkage we usually find in other organisms for inheritance through both the male and female lines. We see here that the majority of gametes produced by the F_1 females carry the parental combinations of alleles. These arise from crossing over (chiasma formation) having taken place in meiosis in some part of the chromosome *other than* between the gene loci for *vg* and *cn* (Fig. 3.11a). Correspondingly, only a minority (10%) of gametes produced are of the recombinant types.

The genes are said to be linked because the parental combination of alleles are associated together in their inheritance in more than 50% of gametes and

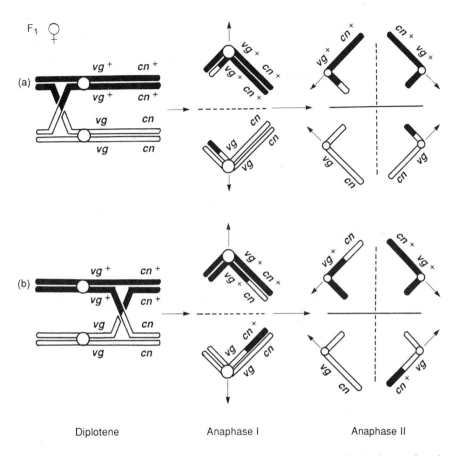

Diplotene   Anaphase I   Anaphase II

Fig. 3.11 Diagram showing segregation of chromosomes and alleles during female meiosis in Drosophila, (a) when there is *no* crossing over between *vg* and *cn*, and (b) when there *is* crossing over between these two loci.

progeny. The recombinant gametes arise when crossing over has taken place between the two gene loci (Fig. 3.11b). Note from Fig. 3.11b that when crossing over does occur between the two gene loci in question it results in two recombinant and two parental type gametes in each meiosis. If crossing over happened between the two gene loci in every meiosis 50% recombination would occur. This is the maximum recombination possible, because crossing over involves individual chromatids, not whole (duplicated) chromosomes. This maximum percentage of recombination would only happen if the two genes were widely separated, such as being located at opposite ends of the chromosome. In the experiment above we have only 10% recombination. This means that the two genes are fairly close together, and that crossing over takes place most of the time in some region of the bivalent other than between these two loci.

Note that had the original cross been set up such that the two mutant traits were carried in different parents (*vg* + x + *cn* , or its reciprocal) the F_2 nonrecombinant and recombinant progeny types would be reversed. Thus the phenotypes *vg* + and + *cn* would be the non-recombinant (parental) types and would represent about 90% of the testcross progeny, while the phenotypes + + and *vg cn* would be the recombinant types and occur in about 10% of cases.

6 Sex Linkage

The demonstration of partial linkage can also be made using sex-linked genes, i.e. genes carried in the X chromosome. The simplest exercise to set up is that involving white eye (*w*) and miniature wing (*m*), with both mutant alleles carried in the female parent, which avoids the need to obtain virgin females at the F_1 stage. Actual results are shown in Table 3.6 and their explanation and interpretation given in Fig. 3.12.

Table 3.6 F_1 and F_2 results from a Drosophila cross involving the sex-linked traits white eye (*w*) and minature wing (*m*).

Parents:	white eye, miniature wing (female) x wild type eye and wild type wing (male), symbolized *w m* x + +.		
F_1	Females wild type (+ +), males white miniature (*w m*), because the genes are in the X chromosome. Allow the F_1 males and females to mate.		
F_2	+ +	375	non-recombinants (parentals)
	+ *m*	151	
			recombinants (32% of total)
	w +	176	
	w m	320	non-recombinants (parentals)
	Total	1022	

The F_1 females are all wild type and heterozygous at both loci, whereas the males are mutant in phenotype and hemizygous. The mating of the F_1 is comparable to a testcross. This is because the males, being hemizygous, contribute only two kinds of gametes, either X-carrying gametes (*w m*), which will go to the daughter progeny, or the Y-carrying ones, which go to the male progeny. In either event, because of the absence of dominant alleles in the male gametes, the phenotypes of the progeny will reveal the genotypes of the gametes coming from the female F_1.

Note that the percentage of recombinant progeny in this exercise (32%) is higher than it is for *vg cn* x + + (10%; see Exercise 5). This means that *vg* and *cn* are closer together (in chromosome II) than are *w* and *m* (in the X chromosome). Extensive genetic maps giving the order and relative distances apart of many genes in Drosophila and other organisms have been constructed using linkage data of this type (Fig. 3.13).

7 Combining a Monohybrid Cross with Sex Linkage

It is possible to combine two (or more) Drosophila exercises into one, thereby saving considerable time. A very useful exercise is to combine monohybrid inher-

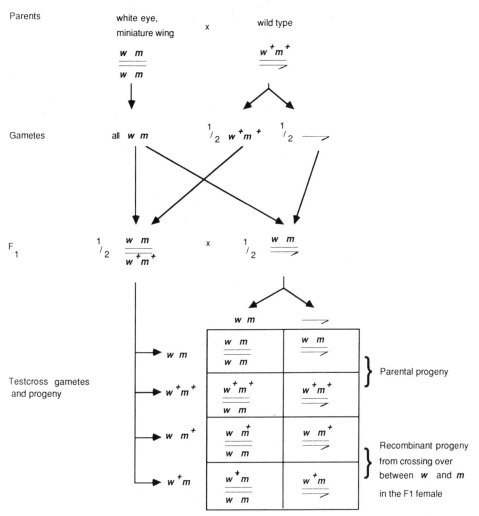

Fig. 3.12 Diagram of the F_1 and testcross progeny of the Drosophila cross white eye, miniature wing x wild type.

itance of *e* (ebony) with sex linkage of *w* and *m*. This is done by mating virgin, white miniature (*w m*) females with ebony bodied (*e*) males, obtaining the F_1, and then allowing these to mate to produce the F_2. Students can be told that they are dealing with ebony, white and miniature, or this information can usefully be withheld until much later. For the latter approach, students first compare body colour, eye colour and wing length in wild type flies and in flies to be used in setting up the cross, with the latter labelled simply as 'parent 1' and 'parent 2'. The procedure in more detail is:

1 Obtain virgin, white eye, miniature wing (*w m*) females and mate them with ebony body (*e*) males. For large classes it is advisable to provide students with virgin females collected for them beforehand.

2 Allow the F_1 to hatch. Examine about 20 F_1 flies, determining their sex and

CHROMOSOME NUMBER

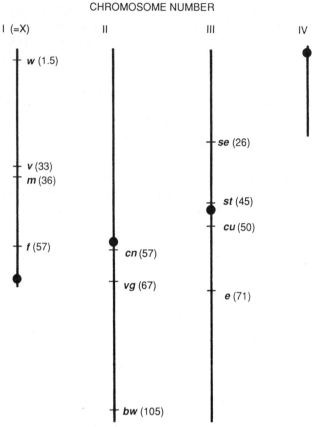

Fig. 3.13 Simplified diagram of the linkage maps of *Drosophila melanogaster*. Map positions are given in brackets. See Table 3.1 for a description of the mutants. • = the centromere.

Table 3.7 F_2 results from the Drosophila cross white eye, miniature wing (female) x ebony body (male); i.e. *w m* x *e*.

Phenotype	Male	Female	Total
w m e	34	43	77
w m +	102	122	224
w + e	18	33	51
w + +	80	89	169
+ m e	25	10	35
+ m +	51	57	108
+ + e	51	55	106
+ + +	181	195	376
Total	542	604	1146

phenotype (body colour, wing size and eye colour). All flies of both sexes will be wild type for body colour, because ebony is not sex-linked and is a recessive trait. But while females are wild type in eye colour and wing size, males are white eyed and miniature winged, because white and miniature are sex-linked recessive traits.

3 Allow the F_1 to mate and produce the F_2. Examine the F_2 flies as before, scoring sufficient to give a class total of about 1000. Typical results (from a class of 36) are given in Table 3.7.

The results are best analysed in two steps, as follows.

(i) *Analysis step 1.* Consider body colour *alone*, ignoring at this stage eye colour and wing size. This is done by combining *all* flies that are wild type in body colour into one group, and all flies that are ebony bodied into another group, males and females separately. The results for the above data are given in Table 3.8.

Table 3.8 Data of Table 3.7 grouped so as to show the pattern of inheritance of ebony (*e*) alone.

	Phenotype		
	+	e	Total
Female	463	141	604
Male	414	128	542
Total	877	269	1146

Note from this table that as for the F_1 there is no relationship between body colour and sex, because ebony is not sex-linked. (The larger number of females derives from the fact that this sex tends to hatch first in young cultures.) Note also that wild type flies are about three times more common than ebony ones, leading to the conclusion that *e* shows a standard monohybrid pattern of inheritance.

(ii) *Analysis step 2.* Now consider eye colour and wing size *alone*, i.e. ignoring body colour. This is done by combining the flies into eight categories, as shown in Table 3.9.

Table 3.9 Data of Table 3.7 grouped to show the pattern of inheritance of white and miniature (*w* and *m*) alone.

	Phenotype				
	+ +	w m	+ m	w +	Total
Female	250	165	67	122	604
Male	232	136	76	98	542
Total	482	301	143	220	1146

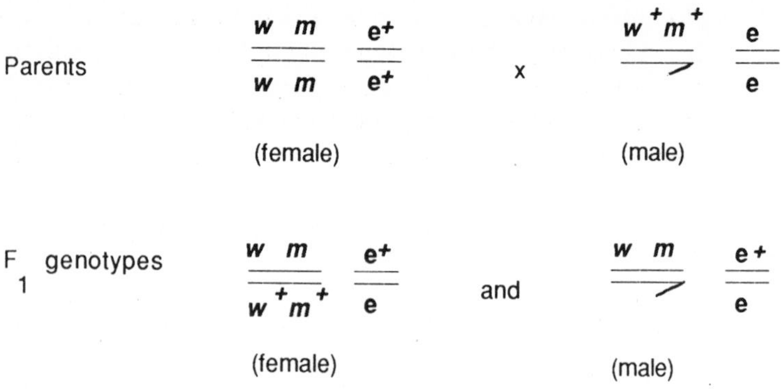

Parents:

$$\frac{w\ m}{w\ m}\ \frac{e^+}{e^+}\ \text{(female)} \quad \times \quad \frac{w^+ m^+}{\diagup}\ \frac{e}{e}\ \text{(male)}$$

F_1 genotypes:

$$\frac{w\ m}{w^+ m^+}\ \frac{e^+}{e}\ \text{(female)} \quad \text{and} \quad \frac{w\ m}{\diagup}\ \frac{e^+}{e}\ \text{(male)}$$

GAMETES PRODUCED BY THE F_1

FEMALE GAMETES	Origin	MALE GAMETES	Origin
$w\ m\ e^+$ $w^+ m^+ e$ $w\ m\ e$ $w^+ m^+ e^+$	By independent segregation of the two pairs of chromosomes	$w\ m\ e^+$ $\diagup\ e$ $w\ m\ e$ $\diagup\ e^+$	By independent segregation of the two pairs of chromosomes
$w\ m^+ e^+$ $w^+ m\ e^+$ $w\ m\ e^+$ $w^+ m\ e^+$	By crossing over between w and m and independent segregation of the two pairs of chromosomes	Not produced (no crossing over in male Drosophila)	Not applicable

The F_2 are obtained by combining each female gamete with each male gamete (32 combinations though some will give like results).

Fig. 3.14 Diagram showing the origin of the F_1 of the Drosophila cross white eye, miniature wing, normal body colour (female) x normal eye, normal wing, ebony body (male); and of the gametes produced by the F_1 as a result of independent segregation of e and $w\ m$ and crossing over between w and m. Note that there is no crossing over in male Drosophila.

Note from Table 3.9 that recombinant progeny (+ m and w +) are present in about the same frequency (32%) as in the $w\ m$ x + + exercise described previously.

4 An attempt should now be made to reconstruct the whole cross, by combining the approaches used for the monohybrid cross (Exercise 1) and the cross with sex linkage (Exercise 6). Here it is important to consider *all three* gene loci at each stage, while taking into account the fact that w and m are on the X chromosome while e is on an autosome, as shown in Fig. 3.14.

8 Drosophila Genetics with 'Unknowns'

Useful and enjoyable exercises with Drosophila can be arranged by presenting students with 'unknown' flies or 'unknown' crosses to work with, thereby giving a more experimental flavour to the practical work.

For studying monohybrid and sex-linked inheritance a very useful pair of 'unknowns' is white-apricot (w^a) and brown eyes (bw). w^a produces pink eyes, is an allele of w and thus has a *sex-linked* pattern of inheritance. The brown eye mutation (bw), however, is an *autosomal recessive* and has a standard monohybrid pattern of inheritance. Cultures of white-apricot and brown-eyed flies are given to students labelled 'unknown 1' and 'unknown 2', along with a culture of wild type flies for comparison and with which they set up appropriate crosses. Students are asked to establish the inheritance pattern of each 'unknown', after considering the theory, setting up appropriate crosses and analysing the results. A simple approach is to tell students beforehand that one unknown involves a gene on chromosome II (for bw) and the other involves a gene on the X chromosome. Alternatively, students may be left entirely in the dark, or perhaps given a 'key to inheritance patterns' with which to work.

After working with w^a as an unknown, an interesting additional exercise involves considering whether the mutation for pink eyes (w^a) is allelic to w (white eyes) or not. The F_1 of a cross between the two mutants will provide the answer, since if they are allelic the mutations will not complement each other and the F_1 progeny will therefore be all mutant. But if they are *not* allelic the F_1 females will be wild type (and the males mutant).

Another very enjoyable approach using 'unknowns', this time involving dihybrid inheritance, is to provide students with F_2 flies of crosses involving eye colour mutants such as scarlet (st), cinnabar (cn), and brown (bw). Students are required to explain the crosses, including parents and F_1. Two such crosses are given below. For each exercise students are told only that the female parent had white eyes (don't say w !) and that the male parent was wild type. The procedure then is for students to examine the F_1, which are wild type, and then to empty these from the bottles and wait for the F_2 to hatch. The F_2 are scored for eye colour over a period of 4–5 days, as they emerge. Males and females are treated separately to check for sex linkage. The extended period of observation reduces any inaccuracy which may arise from differential viability of the mutant phenotypes. Actual results are shown in Table 3.10. There are four phenotypic classes in Table 3.10, which occur in a ratio of 9 : 3 : 3 : 1, for both males and females. This ratio in the F_2 tells us that two autosomal recessive alleles are segregating independently from a heterozygous F_1. The white eyed female parent must therefore have been homozygous for two different recessive genes that interact to give the white phenotype of one parent and the white eye class in the F_2. The other two mutant classes in the F_2 tell us that one of the two recessive genes produces

Table 3.10 F_2 results from an 'unknown' Drosophila cross involving eye colour.

	wild type (dull red)	Eye colour phenotype brownish	bright red	white	Total
1st scoring	497	204	164	40	905
2nd scoring	196	56	61	32	345
Total	693	260	225	72	1250

χ^2 (9 : 3 : 3 : 1) = 3.85, $d.f.$ = 3, P = 0.3, not significant.

brown eyes, the other bright red eyes.

The deduction, therefore, is that white derives from an absence of both red and brown colour; and that brown and red combine to give the dull red wild type colour.

Reference to the linkage map of Drosophila (Fig. 3.13) will indicate that a number of pairs of unlinked genes will give the required result, such as scarlet (*st*) and brown (*bw*), and cinnabar (*cn*) and sepia (*se*). The cross involving *st* and *bw* is illustrated in Fig. 3.15.

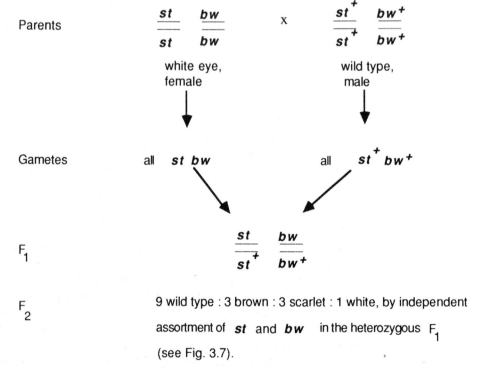

Parents

white eye, female

X

wild type, male

Gametes   all *st bw*   all *st*⁺ *bw*⁺

F_1

F_2 9 wild type : 3 brown : 3 scarlet : 1 white, by independent assortment of **st** and **bw** in the heterozygous F_1 (see Fig. 3.7).

Fig. 3.15 Explanation of the 'unknown 1' Drosophila cross involving the unlinked genes *st* and *bw*, which produce a 9 : 3 : 3 : 1 ratio in the F_2 by independent segregation.

The value of this exercise lies in its illustration of the way in which two different genes interact to produce one character (eye colour), and in serving as a useful introduction to the biochemical genetics of eye pigment formation in Drosophila. Thus after the above exercise students can be referred to the two biochemical pathways to pigment production in Drosophila, namely for the pterins (red) and ommochromes (brown). Bright red eyed flies (st or cn or v) carry a mutation in one of the genes involved in the production of brown pigment. They produce no brown pigment and thus have bright red eyes by default. Similarly, brown eyes (bw or se) each derive from a mutation in genes involved in the production of red pigment. Note that white eyed flies of the genotype st/st, bw/bw are genetically quite distinct from those carrying the white eye mutation w. Mutation in the w locus inhibits the deposition of pigment in the eye, even though pigment is being produced.

Having worked out and understood 'unknown 1' students will then be in a position to attempt 'unknown 2' (Table 3.11).

Table 3.11 F_2 results from a second 'unknown' Drosophila cross involving eye colour.

	wild type (dull red)	Eye colour phenotype brownish	bright red	white	Total
1st scoring	533	92	88	62	775
2nd scoring	51	12	22	28	113
Total	584	104	110	90	888

χ^2 (5 : 1 : 1 : 1) = 5.94, d.f. = 3, P = 0.1, not significant.

In Table 3.11 there are again four phenotypes in the F_2, but in this case they approximate best to the unusual ratio of 5 : 1 : 1 : 1.

The key to unravelling this problem lies in recognising that the four phenotypes suggest a dihybrid F_2 ratio. The numerals 5 : 1 : 1 : 1 add to 8 (= 4 x 2), which is half of the 16 (4 x 4) that the numerals of the normal dihybrid ratio (9 : 3 : 3 : 1) add to. This suggests that the F_1 is, indeed, heterozygous for two recessive alleles which appear to segregate independently, as in a standard dihybrid cross, but that for one of the sexes there are two genotypic classes of gametes missing. As before it is evident also that two genes interact in the original female parent, and in one of the F_2 classes, to give white eyes. The explanation for the absence of two gametic classes is that the genes are linked and that there is no crossing over in the male. This latter point may be hinted at if students find the solution to this unknown difficult. The apparent independent segregation which occurs in females arises because the two genes are widely spaced in the chromosome concerned and thus there is crossing over between them in *every* meiosis. Reference to the linkage map (Fig. 3.13) will indicate that the appropriate gene pair is brown (bw) and cinnabar (cn). Cinnabar is nearly 50 map units away from bw in chromosome II. The interpretation of the cross is given in Fig. 3.16.

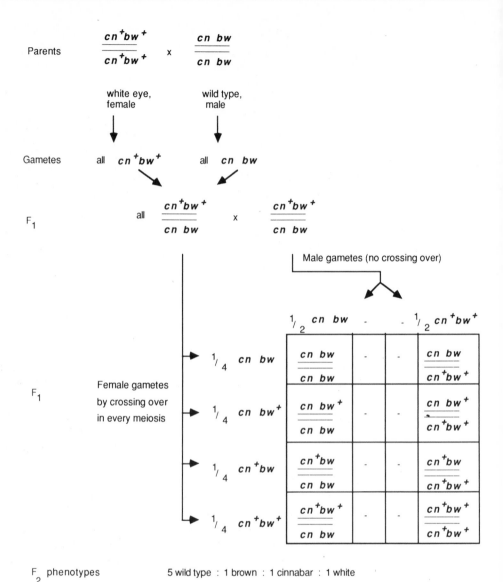

Fig. 3.16 Explanation of the 'unknown 2' Drosophila cross involving the linked genes *cn* and *bw*, which produce the unusual F_2 ratio of 5 : 1 : 1 : 1 because the gene loci are about 50 map units apart, and because there is no crossing over in male Drosophila.

9 Environmental Effects on Gene Expression

Most aspects of the phenotype are influenced by both the genetic constitution of the organism concerned and by the environment in which it is reared. We say that the phenotype is the product of interactions between the genotype and the environment. A convenient experiment which demonstrates how the action of a single gene can be modified by the environment is that involving curled wing (*cu*) in Drosophila. The expression of *cu* is sensitive to temperature. Flies homo-

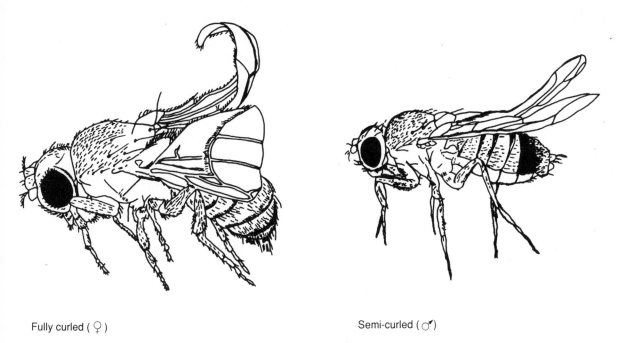

Fully curled (♀)

Semi-curled (♂)

Fig. 3.17 Curled and semi-curled wing phenotypes in male and female Drosophila. Despite their different degrees of curliness both flies are homozygous for the *cu* allele.

zygous for curled wings (*cu/cu*) that are reared at high temperature have wings that are curled up at the tips (Fig. 3.17). But when flies of this same genotype are reared at low temperature a proportion of them show straight wings. Thus we say that the *penetrance* of the *cu* allele is incomplete. Also, flies reared especially at low temperatures have wings that vary in their degree of curliness. Thus we say that *expressivity* of the gene is variable. Penetrance and expressivity of the *cu* allele vary with sex.

In the following experiment, homozygous (*cu/cu*) flies are reared at three different temperatures, and then analysed for penetrance and expressivity of the curled wing character. The experimental procedure is as follows.

1 Set up three cultures of *cu* flies (genotype *cu/cu*) from stock bottles. The females do not have to be virgin. Label the cultures 20°C, 25°C and 28°C. Incubate all three cultures at 25°C initially, until just before pupation begins (this speeds up the experiment; and temperature has its effect on the expression of *cu* only just before hatching).

2 Remove the adults from each culture and transfer the cultures to their respective temperatures. For 25°C and 28°C this normally requires an incubator set and stabilized at these temperatures. For 20°C, a cool room where the temperature does not fluctuate much outside 18–20°C is satisfactory, though a cooling incubator (refrigerator!) set and stabilized at 20°C is the ideal. The 20°C culture will take about 3 weeks to produce adult flies.

3 Etherize the flies in convenient batches as they hatch and then classify males and females separately for wing form. Use three categories, namely curled,

semi-curled and non-curled (Fig. 3.17), using your own guidelines. There will be some flies that are difficult to score clearly, but these will not affect the overall result substantially. The important point is to be consistent in the scoring of different batches.

4 Tabulate the data and then calculate penetrance and expressivity for each sex (Table 3.12), using the following formulae:

$$\text{Penetrance} = \frac{\text{no. of affected flies (curled and semi-curled)}}{\text{Total number of flies}} \times 100$$

$$\text{Expressivity} = \frac{\text{(no. of curled flies} \times 2) + (\text{no. of semi-curled} \times 1)}{\text{Total number of } penetrant \text{ flies} \times 2} \times 100$$

Note that in the calculation for expressivity a weighting function is used, which scores two for each curled fly, and one for each semi-curled fly, thus enabling the results to be combined into one overall figure of expressivity.

Table 3.12 Numbers of curled, semi-curled and non-curled (straight) winged Drosophila flies of genotype *cu/cu* at each of three temperature regimes, together with calculated values of penetrance and expressivity.

	20°C		25°C		28°C	
	Female	*Male*	*Female*	*Male*	*Female*	*Male*
Curled	0	0	40	31	268	220
Semi-curled	60	42	314	290	367	351
Non-curled	54	81	150	201	80	91
Total	114	123	504	522	715	662
Penetrance (%)	52.6	34.2	70.2	61.5	88.8	86.3
Expressivity (%)	50.0	50.0	55.7	54.8	71.1	69.3

The results can now be shown in the form of histograms (Fig. 3.18).

The effects of temperature on the *cu* gene can be discussed in terms of the three-dimensional structure and functional capacity of proteins.

As a test of how well students grasp the significance of this experiment it is useful to ask them to write down the *genotype* of each of the curled, semi-curled and non-curled flies in their 28°C culture. The answer, of course, is *cu/cu* (homozygous *cu*) for *all three* phenotypes, *not cu/cu , cu/cu⁺* and *cu⁺/cu⁺*, respectively, as many will suppose. If students fall into this error they should be directed again to the fact that the flies initially used to set up their cultures were all *cu/cu*.

Another way of testing student understanding is as follows. 'Considering both genotype and phenotype (penetrance only), predict the F_1 of a cross between two non-curled flies obtained from your 20°C culture, this cross being reared at 28°C.'

The answer requires students first to appreciate the fact that the genotype of the parents (and F_1) is *cu/cu* (not *cu⁺/cu⁺*); and requires them to use the results

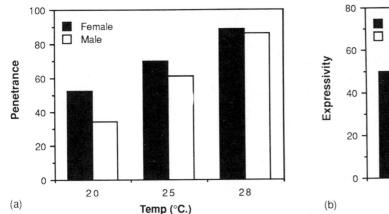
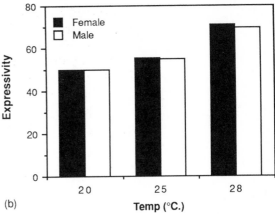

(a)   (b)

Fig. 3.18 Histograms showing (a) penetrance and (b) expressivity for the *cu* allele in male and female Drosophila reared under three temperature regimes.

obtained from their 28°C experiment to predict the distribution of phenotypes in the F_1, which will include curled and semi-curled flies as well as non-curled ones.

10 Genetics of a Quantitative Trait

So far in this chapter we have been dealing with *qualitative (discontinuous)* traits, such as vestigial versus normal wing, white versus normal eye colour. For these traits the phenotype is one extreme or the other, with no or few intermediate conditions. Most traits studied in a laboratory situation are of this type. The variation we see in nature, and the variation of most importance in plant and animal breeding, however, is of the *quantitative (continuous)* type, such as variation in height, weight, growth rate, productivity, colour, etc. For these traits there is a spectrum of phenotypes ranging from one extreme to the other. Such traits are strongly influenced by both genetic and environmental factors. The genes concerned are called *polygenes*, because there are *many* of them affecting the same phenotypic trait. They act by each contributing usually a small, equal amount to the phenotype, so that the degree to which the trait is expressed depends on the number of contributing genes carried by the organism. The following exercise illustrates how a quantitative trait can be studied and how genetic and environmental factors contributing to it can be distinguished. The exercise concerns two inbred lines of Drosophila exhibiting high and low numbers of sternopleural bristles.

Wild type Drosophila vary in the number of bristles (chaetae) found in certain parts of the body, such as the sternopleura. The sternopleura are a pair of roughly triangular skeletal plates found on the underside of the body extending from the midline between the mid and fore legs round towards the sides of the body (Fig. 3.19). The bristles on each plate comprise two or three long major bristles nearest the body margin and a rather regular row of small bristles extending towards the body midline.

To observe and count the sternopleural bristles it is necessary to etherize and place each fly on its back, slightly tilted to one side. Illuminate the fly well (from above) and examine it under a dissecting microscope at about x40 magnification.

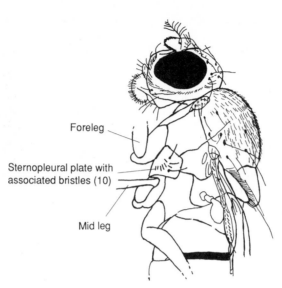

Fig. 3.19   Diagram showing a sternopleural plate and its associated bristles (three large and seven small) on one side of a Drosophila fly.

Students may find it difficult initially to identify the sternopleura and their bristles amongst what seems at first to be a maze of skeletal plates and bristles. But just a little bit of careful study with reference to Fig. 3.19 will soon enable them to find the appropriate bristles and count them. They should not confuse bristles on the base of the legs and close to the body midline with the sternopleural bristles. Males, being smaller, have fewer bristles than females. This difference may be an aspect of study by a class; or may be avoided by examining only one sex (conventionally the female).

Cultures of inbred wild type lines of Drosophila exhibiting 'high' and 'low' sternopleural bristle numbers can be obtained from usual sources. Ideally the high line should have about 20 bristles per pair of sternopleural plates, and the low line about half this number. Many more than 20 leads to difficulties in counting the bristles accurately; and many fewer than 10 limits the amount of variability.

Special care shoul be exercised in maintaining 'high' and 'low' stocks, for it is not easy to re-isolate the original strain from a contaminated culture. Labelling one parental stock in red, and the other in blue, for example, helps to avoid confusion during subculturing. For student use, stocks are best labelled in a way that avoids disclosing beforehand their relative numbers of bristles.

Experimental approaches

The exercise presented here can be performed at various levels of sophistication. In its simplest form, students are provided with files of each of the high and low lines, reared for this purpose in one incubator (to avoid consistent differences in culture conditions for the two strains). Students then count the number of bristles in a sample of flies from each of the high and low lines. About 5 – 10 flies are sufficient in a large class, though with a small class (< 20) this number should be increased so as to give adequate data for analysis. The bristles on *both* sterno-

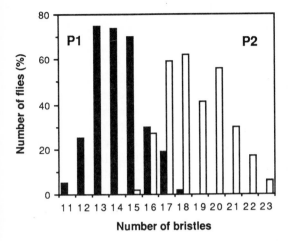

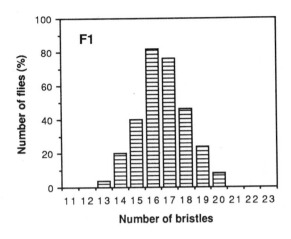

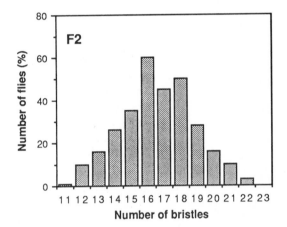

Fig. 3.20 Histograms showing sternopleural bristle numbers in two inbred lines of Drosophila (P1, solid bars and P2, open bars), and in their F_1 and F_2 progeny.

pleural plates should be counted, with individual counts being entered carefully onto tally sheets. It is advisable to have all students counting flies from the *low* line first, and under close supervision initially, by the end of which time they will be sufficiently experienced to tackle the slightly more difficult high line flies. Ensuring that students count exactly the same number of flies makes data handling much easier than otherwise. Individual results should be pooled as a frequency distribution (the number of flies in each bristle number category), from which the mean bristle number for each line can be calculated.

Actual results from a class of 30 first year university students are shown in the upper two rows of Table 3.13 and in the upper (composite) histogram of Fig. 3.20. Note the *continuous* nature of the phenotype, in that a full range of bristle numbers (10 – 23) is found, not just high and low numbers. The distribution of bristle numbers in each of the two lines is 'bell shaped', with about as many observations above the mean as below it. This is typical of the distribution pattern of a quantitative trait.

The variation in bristle number *within each line*, given by the spread of values on each side of the mean, derives from *environmental* effects on bristle number. This is because the flies within each line are strongly *inbred* (i.e. derive from repeated brother–sister matings) and thus are genetically all the same and are all homozygous for their bristle number genes. Thus the variation in bristle number within each line must be environmental in origin, deriving from small differences in culture conditions (food quality, temperature, etc.).

Although the bristle number distributions of the two lines overlap they are clearly very different. The difference, summarized by the different means (Table 3.13), cannot be environmental in origin, since the cultures were reared under the one broad set of environmental conditions. The *difference between the two lines*, therefore, is an indication of *genetic* differences in bristle number, i.e. differences in the number of polygenes contributing to bristle number. Thus the genotypes of the low and high lines may be represented **AAB'B'C'C'D'D'E'E'** .... and **AABBCCDDE'E'** ..., respectively, for example, where **A,B,C** and **D** .... contribute equally and additively to bristle number, but **B', C',D'** and **E'** .... have no effect.

The above exercise may be followed by consideration of bristle numbers expected for the F_1 and F_2 generations formed by crossing high and low lines. Alternatively, the practical work itself can be extended to include students producing these two generations in the standard way (or being provided with them). Bristle numbers can then be obtained for all three (parental, F_1 and F_2) generations. From the data obtained an estimate of the *heritability* (the contribution that genes make to variation) of bristle number can be obtained. For this purpose the simplest approach is for bristle numbers to be recorded for the parents, F_1 and F_2 generations sequentially, as the experiment proceeds. Data obtained in this way are also shown in Table 3.13 and Fig. 3.20. They are analysed and interpreted in the following way.

Table 3.13 Frequency distribution, mean and variance for sternopleural bristle number in each of two inbred lines of Drosophila having 'high' and 'low' bristle numbers; and in their F_1 and F_2 generations. P = Parent.

Strain and generation	Number of bristles													Mean	Variance
	11	12	13	14	15	16	17	18	19	20	21	22	23		
'Low' (P1)	5	25	75	74	70	30	19	2						14.2	2.02
'High' (P2)					2	27	59	62	41	56	30	17	6	18.7	3.27
F_1 (P1 x P2)			4	20	40	82	76	46	24	8				16.6	2.18
F_2 (F1 x F1)	1	10	16	26	35	60	45	50	28	16	10	3		16.6	4.92

1   *The mean* ($\overline{X}$). This is obtained by summing the number of bristles and then dividing by the number of flies. From the frequency distribution this is done by multiplying the number of flies in each bristle number category by its bristle number, summing these values and then dividing by the total number of flies, i.e.

$$\overline{X} = \{f(X_1) + f(X_2) + f(X_3) \ldots + f(X_N)\} / N$$
where f = frequency
$X_1 \ldots X_N$ = the number of bristles in each bristle number category
N = the total number of flies and
$/$ = divided by.

2 *The variance.* This is symbolized V. It is a measure of the distribution or spread of values round the mean, and thus a measure of the variability of bristle number. To obtain the variance, each value is related to the mean by subtraction, the values so obtained are squared (to get rid of negatives), and the squared values summed. The sum of the squared values (sums of squares) are then divided by the number of observations minus one ($N - 1$, or the 'degrees of freedom'). Thus,

$$V = \{f(X_1 - \bar{X})^2 + f(X_2 - \bar{X})^2 + f(X_3 - \bar{X})^2 \ldots + f(X_N - \bar{X})^2\} / N - 1$$

Note that when using the above formula squaring must be done *before* multiplying by f. The division is done because the sums of squares will vary according to the number of observations. $N - 1$ is used as the divisor (rather than N) because if there were only one observation the mean would be the same as this observation and there would be no variance.

3 *Variance due to environmental differences.* As indicated above, the two parent lines are each strongly inbred and thus genetically very homogeneous and homozygous. Therefore the spread of bristle number round the mean for each parent generation (i.e. V_{P1} and V_{P2}) must be due entirely to environmental factors influencing bristle number during fly development. Variance due to environmental effects is symbolised V_E. Thus,

$$V_{P1} = V_{P2} = V_E$$

The F_1 generation will also be genetically homogeneous, but will be *heterozygous* for those genes for which the two parents differ. Thus for the example given above the F_1 will all be **AABB'CC'DD'E'E'**.

Therefore the variance of the F_1 is also due solely to environmental factors. Thus,

$$V_{F_1} = V_E$$

Hence the variances of the F_1 and parent generations give us three independent measures of V_E. In the calculations below we have used the average of the three (Table 3.13) as our measure of V_E (i.e. 2.49).

It will be seen that the distribution of the F_1 is approximately mid-way between those of the two parents (Fig. 3.20). This indicates that the alleles for bristle number for which the two parents differ are not showing dominance or recessiveness amongst themselves (and thus are symbolized **A** and **A'**, **B** and **B'** etc. rather than **A** and **a**, **B** and **b**).

4 *Variance due to genetic differences.* We see that the F_2 variance is much greater than that of either the parent or F_1 generations. It spans nearly the whole spectrum of bristle numbers shown by the parents combined. This is because not only is there environmental variation contained therein, but also genetic variation. Genetic variance (symbolized V_G) derives from the fact that there will have been segregation and recombination of polygenes during meiosis in the F_1, which will produce many genotypes in the F_2 ranging from one extreme (produced rarely) through all intermediates to the other extreme (also produced rarely).

Thus,

$$V_{F_2} = V_G + V_E$$

If we assume that the F_2 received the same culture conditions as the parental and F_1 generations then we can obtain a measure of V_G by subtracting V_E from the F_2 variance.

For the data of Table 3.13,

V_{F_2} = 4.92 and, therefore,
V_G = 4.92 - 2.49
    = 2.43

5 *Heritability* (symbolized h^2). Heritability (in the broad sense used here) is the proportion of the total variance that is accounted for by genetic variation. It is given, therefore, as V_G/V_{F_2}, and will vary from 0 (variability due entirely to environmental differences) to 1 (variability due entirely to genetic differences).

For the data of Table 3.13,

h^2 = 2.43 / 4.92
   = 0.49

In other words, appoximately 50% of the variability in sternopleural bristle number in these lines of Drosophila is due to genetic factors, 50% due to environmental factors.

The straightforward approach used above can introduce a significant amount of artificial (extraneous) variation into the exercise, which can significantly affect the heritability value obtained. This artificial variation derives partly from the fact that students can vary in the accuracy with which they score bristles, and partly because different growth conditions may apply to different generations of flies. Therefore the ideal way of conducting the above exercise is to set up *dual* crosses in such a way that all cultures (parents, F_1 and F_2) of the whole class can be *positioned randomly* on one shelf of an incubator and raised together *as a single experiment*. This enables the analysis of bristle numbers to be done on flies of the same age, reared under the same growth conditions; and enables an estimate of extraneous variation to be made.

This ideal method of conducting the exercise requires a lot of careful planning and execution and is thus recommended only for small, advanced classes. The interested reader is referred to the excellent article by Lawrence and Jinks (1973) for further details. Our advice for junior classes is to set up the exercise in a way that best suits a particular classroom situation. Remember always that the purpose of the exercise is not to rediscover the precise heritability of sternopleural bristle number in Drosophila, but rather to introduce students to some of the procedures used in the analysis of quantitative traits. A simplified experiment, followed by class discussion of its limitations, gives students a very worthwhile introduction to quantitative genetics.

REFERENCES

Demerec M. and Kaufmann B.P. (1973). *Drosophila – A Guide*. 8th Edition. Washington, Carnegie Institute of Washington.

Haskell G. (1961). *Practical Heredity with Drosophila*. Edinburgh and London, Oliver and Boyd.

Lawrence M.J. and Jinks J.L. (1973). Quantitative genetics. In *Practical Genetics*, ed. P.M. Sheppard; chp. 3. Oxford, Blackwell Scientific Publications.

Strickberger M.W. (1962). *Experiments in Genetics with Drosophila*. New York, Wiley and Sons.

Chapter 4   GENETICS WITH YEAST

The exercises described in this chapter are suitable for all levels of study from senior school to university undergraduate. The yeast species used is the one commonly employed in brewing and baking. It poses no biological hazards. Standard procedures in the safe handling of micro-organisms (see Appendix) should be used, and the normal precaution of disposing of plates by autoclaving taken. A bunsen flame is used to sterilize instruments employed in handling yeast, and thus this work should not be done at the same time as anaesthetizing Drosophila flies with ether. For the treatment of yeast cells with ultraviolet light (Exercise 3) it is sensible to wear rubber gloves and safety glasses during the irradiation procedure so as to avoid damage to eyes and skin.

INTRODUCTION AND BACKGROUND

Yeasts, of which there are many different types, are fungi. They are unusual in that they do not have a true mycelium but instead are unicellular. Thus they grow more like bacteria than like a typical fungus. The growth characteristics of yeasts, and the simple ways in which they can be cultured and manipulated experimentally, make them ideal organisms for genetic studies. In this chapter we describe exercises with bread yeast, *Saccharomyces cerevisiae*, to demonstrate (i) the relationship between genes, enzymes, biochemical pathways and the phenotype under normal and mutant situations; (ii) the phenomenon of genetic complementation; and (iii) the use of ultraviolet light for the induction of mutations.

Bread yeast (simplified to just 'yeast' hereafter) grows naturally on substrates that are rich in sugar such as ripe fruits and nectar. In the laboratory it grows readily on an agar based medium containing a carbon and nitrogen food source, from which the organism will manufacture all its cellular requirements.

Yeast is an ascomycete fungus, reproducing sexually by forming ascospores in a sac-like structure called an ascus. Yeast cells also reproduce asexually by 'budding' by mitosis. The life cycle is illustrated in Fig. 4.1. There are two mating types, designated *a* and α, which are determined by a single gene with two alleles. When kept apart these mating types will grow as haploid strains, reproducing asexually. But when mixed, cells of opposite mating types readily fuse in pairs to form diploid cells. These diploid cells will also reproduce asexually by budding. But when grown on a 'starvation' medium diploid cells undergo meiosis to produce asci, each containing four haploid ascospores. During ascospore formation the mating type alleles segregate to produce *a* and α type spores, from which the single mating type haploid strains are reformed.

Yeast grows rapidly when cells are spread onto a yeast minimal agar medium and kept for a few days at room temperature. Yeast minimal medium (YMM) is a

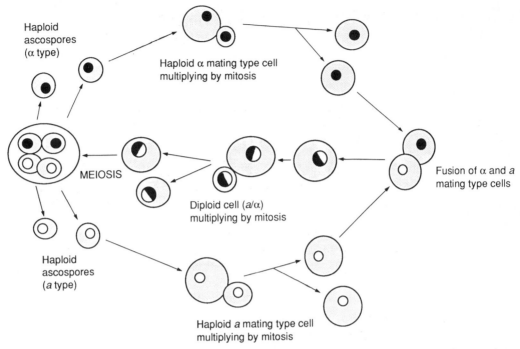

Haploid
ascospores
(α type)

Haploid α mating type cell
multiplying by mitosis

MEIOSIS

Fusion of α and a
mating type cells

Diploid cell (a/α)
multiplying by mitosis

Haploid
ascospores
(a type)

Haploid a mating type cell
multiplying by mitosis

Fig. 4.1 Diagram of the life cycle of the bread yeast, *Saccharomyces cereviseae*. See text for explanation. The two mating types are distinguished by open and filled nuclei, with the diploid cell carrying both relevant alleles.

simple food containing glucose (as a source of carbohydrate) and a supply of nitrogen. With these simple ingredients the normal (wild type) form of the fungus can carry out all the necessary biochemical processes required for its growth and reproduction, growing as numerous white colonies which may merge together to form a dense white growth on the surface of the agar (Fig. 4.2).

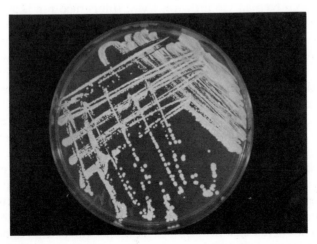

Fig. 4.2 A yeast inoculum streaked out in such a way as to produce colonies which have grown from single cells.

(a)

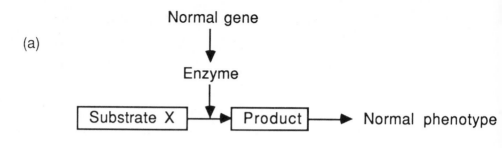

(b)

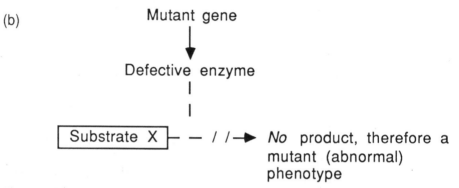

Fig. 4.3 Diagram showing the relationship between genes and the phenotype in biochemical terms, (a) for the normal situation and (b) for the situation in which a mutant gene produces a defective, non-functional enzyme.

Mutant strains of yeast occur that are unable to grow on YMM because each lacks the capacity to produce a particular substance (amino acid, growth hormone, etc.). These mutants arise spontaneously or they may be induced experimentally with chemical mutagens or irradiation with ultraviolet light, etc. A mutant can be made to grow on YMM if this medium is supplemented with the particular substance the yeast is unable to make for itself. It will also grow on a yeast complete medium (YCM), which contains an extract of yeast and thus has all the organic compounds needed for the growth of both wild type and mutant yeasts.

Mutant strains usually have a defect in an enzyme involved in a particular biochemical pathway, and as such they provide an excellent illustration of the way in which genes work to determine phenotypic traits. Fig. 4.3 shows how we can represent the relationship between genes, enzymes, biochemical pathways and the phenotype.

In Fig. 4.3a of this simple pathway, the wild type gene codes for an enzyme which catalyses the conversion of a substrate (X) to a particular product, which contributes directly or indirectly to the phenotype. When the gene is a mutant one (Fig. 4.3b) the enzyme is defective and thus the substrate is not metabolized. The consequent block in the biochemical pathway leads to both an accumulation of the substrate and to the non-production of the product. One or other or both of these consequences leads to an abnormal (mutant) phenotype.

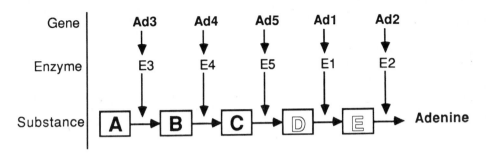

Fig. 4.4 Diagram of the biochemical pathway for the synthesis of adenine in yeast, and the genes and enzymes involved. Precursor substances A–C are white (colourless), while D and E are red.

The exercises to be described in this chapter involve the adenine requiring mutants of yeast, i.e. the so called ad⁻ (ad minus) strains. These mutants are particularly useful because some of them accumulate substrates which are *red* in colour, leading to phenotypes which are visibly distinguishable from the normal white form.

The ad⁻ strains will not grow on YMM. They will, however, grow on YCM or on YMM to which adenine has been added as a supplement. Adenine is one of the building blocks of nucleic acids (DNA and RNA) and its synthesis is controlled by a number of genes in a biochemical pathway of which part is shown diagrammatically in Fig. 4.4.

In the wild type strain there is a normal allele at each of the five gene loci shown, so that all of the enzymes are present and function to catalyse the whole pathway through to adenine. But if a gene is mutant the enzyme controlled by that gene may be absent or may fail to work. Cells carrying such a mutation cannot then synthesize adenine and, because there is no feedback inhibition to turn the whole adenine pathway off, there is a build up of the precursor substance at the place in the sequence before the block. Mutations in genes **ad1** and **ad2** are especially interesting because they cause accumulation of substance D or substance E. These two substances, aminoimidazole ribonucleotide (D for the **ad1** mutant) and aminoimidazole carboxylic acid ribonucleotide (E for the **ad2** mutant), are *red* and produce red coloured yeasts, which also require adenine to be supplied to them for their growth. Mutations in any of the other genes also give rise to adenine requiring strains, but since substances A, B and C are not coloured these mutants are white and cannot be visibly distinguished from the wild type strain except for their inability to grow on YMM.

Several independently isolated mutants at each of the five gene loci of Fig. 4.4, have been isolated at various times. These mutants (e.g. **ad2.0** , **ad2.1**, **ad2.4** and **ad2.6** for the **ad2** locus) represent different allelic forms of a single gene. Multiple alleles of this kind occur because independent mutational events cause changes in different regions of the gene concerned. In the exercises to be described below we show how it is possible to determine whether two independently isolated ad⁻ mutants having identical visible phenotypes belong to the same gene (for example, **ad2.1** and **ad2.4**) or to different genes (for example, **ad1** and **ad2**).

PROCEDURES

General Technique

YCM is highly nutritious and readily picks up contaminants in the form of fungal spores or bacteria from the air, hands and other parts of the body and from instruments. It is essential, therefore, to use sterile procedures and to maintain a good standard of hygiene when the following exercises and experiments are conducted, and to use previously sterilized growth media, culture plates, glassware, etc., as in all microbiological work. Sterilize pasteur pipettes (in small bundles wrapped in paper), glass containers, growth media, etc. in an autoclave (or pressure cooker); and sterilize glass spreaders by dipping them in 95% alcohol and then flaming them, repeating the procedure at least twice. Sterilize a wire loop by heating it to red hot in a bunsen flame and then cooling the loop on the agar of the plate to be inoculated.

Inoculation of Agar Plates

All that is required to grow yeast is to pick up on a sterile wire loop a small quantity of cells (an inoculum) from another culture and to streak the cells back and forth across a sterile plate of YCM. The lid of the agar plate to be inoculated should be only *half lifted* or, if removed completely, should be placed with the inside *face down* on the bench. Also, avoid breathing unnecessarily over the exposed agar plate or working in the presence of draughts. After inoculating a plate, 5 days growth at normal room temperatures produce colonies of yeast cells which will eventually merge together to produce a mass of cells as those shown in Fig. 4.2.

Purifying Cultures

Cultures of yeast sometimes need to be purified (rid of spontaneously formed mutants or other contaminants, for example), at the start of an exercise or when setting up new stock cultures for holding until the next round of exercises. To purify a culture cells must be streaked onto YCM in a particular way so as to be able to re-isolate a strain from a colony derived from a single cell. This process involves streaking cells across one side of the agar plate in a series of parallel lines to thin out the inoculum progressively. The loop is then flamed and the first track of cells are thinned out further by spreading through them *once* at right angles and then thinning out what cells have been picked up, as before. This procedure is repeated a second and then a third time, at which stage the inoculum has become sufficiently dilute to yield individual colonies which have grown from single cells (Fig. 4.2). One of these colonies can then be picked off with a loop and used to grow the purified strain, or to inoculate a stock culture bottle.

Stock Cultures

To set up stock cultures, streak cells from a colony derived from a single cell onto a YCM slant in a universal container. The screw cap of the container should be left loose for 3–4 days until the culture shows good growth (Fig. 4.5), but then

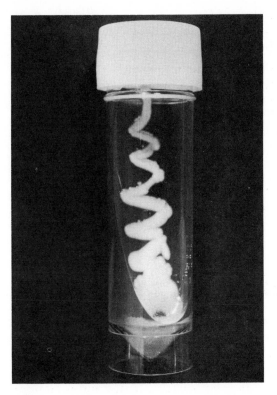

Fig. 4.5 A stock culture of yeast growing on an agar slant of YCM in a universal container.

should be tightened and the container stored in a refrigerator at about 4°C. In this state the stocks will remain viable for at least 1 year, though it is advisable to check them for purity before embarking on a series of experiments.

Refer to the Appendix for sources of stock cultures.

Cell Suspensions

In some exercises it is convenient to have the yeast as a cell suspension in physiological saline. To prepare such a suspension an appropriate amount of yeast is scraped off a plate with a sterile loop or spatula and transferred by shaking into the prepared saline in a sterile bijou bottle. Physiological saline is a 0.85% solution of sodium chloride in water, which is autoclaved for 15 min at 15 p.s.i. before use. Cells suspended in saline will remain viable for several days.

Inoculating Plates from a Cell Suspension

First shake the bottle vigorously to disperse the cells evenly through the saline. *This is very important.* Remove the lid and take up the required quantity of suspension into a sterile pasteur pipette or a sterile loop. A drop or more of suspension is now placed on the surface of an agar plate and then spread over the surface of the agar using a sterile glass spreader.

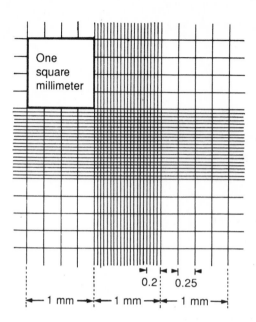

One
square
millimeter

0.2   0.25

←— 1 mm —→←— 1 mm —→←— 1 mm —→

Fig. 4.6 Diagram of the grid markings on a haemocytometer slide. These markings, and the cells being counted, are seen under the microscope, so that the areas for counting are clearly defined.

Estimating Cell Concentrations with a Haemocytometer

In some exercises it is necessary to estimate the concentration of yeast cells in a suspension. The haemocytometer slide is a precision instrument designed to enable this to be done. Concentration is normally expressed as the number of cells per ml (cm^3). The procedure is:

1 Shake the suspension.
2 Moisten the suspending bars on the haemocytometer and apply the coverslip with sliding pressure, so that it holds firmly to the slide.
3 Take a sample of the suspension in a sterile pasteur pipette and allow some of it to run under the haemocytometer coverslip until the enclosed space is filled. The liquid should not run into the channel on either side of the depression.
4 Place the haemocytometer on a microscope stage and focus to observe the grid markings that are etched onto the slide in the central part of the well. The pattern of grid markings is shown in Fig. 4.6. The yeast cells will be visible at the same time.
5 Count the number of cells in each of four 1 mm x 1 mm corner squares. Sum these counts and divide by 4 to obtain the average per square mm. To determine the cell concentration multiply the sq. mm count by 10 to convert to cubic mm (if the chamber is 0.1 mm deep) and then by 1000 to convert to cm^3. If the haemocytometer is one in which the chamber is 0.2 mm deep, multiply the sq. mm count by 5, instead of 10, and then by 1000. Chamber depth will be indicated on the haemocytometer or on accompanying information.

Culture Media

Yeast complete medium (YCM)

Peptone	10 g
Yeast extract	10 g
Glucose	20 g
Agar (oxoid no.3)	15 g
Water	1000 ml

Mix together and autoclave (or pressure cook) at 15 p.s.i. for 15 min and pour into sterile petri dishes or universal bottles. Place the latter on an angle as they cool to form agar slants.

Yeast minimal medium (YMM)

Difco yeast nitrogen base	7 g
Glucose	40 g
Agar	15 g
Water	1000 ml

Prepare and autoclave as above.

Supplemented medium (YMM + adenine)

Make a stock solution of adenine at a concentration of 1.0 mg per ml of distilled water. Autoclave at 10 p.s.i. for 10 min. Meanwhile, make up YMM as above and autoclave separately from the adenine solution. Before the agar solidifies, add 3 ml adenine solution per 200 ml of YMM. Make plates as above.

1000 ml of medium (any of the above three) is sufficient to make approximately 40 plates if these are poured to about $\frac{1}{3}$ depth of the petri dishes.

EXERCISES

1 Phenotypes of Wild Type and Adenine Requiring Yeasts

As a introductory exercise each student can grow the wild type and ad⁻ strains on YCM, YMM and YMM+adenine plates to observe their visual and biochemical phenotypes. Each plate can be marked into halves or quarters with a marker pen, labelled and then wild type and ad⁻ strains streaked onto appropriate sectors. The ideal is to test wild type, **ad1**, **ad2** and **ad3** (or **ad4** or **ad5**), giving four different strains on each of three plates. A simplified experiment is to test only wild type and **ad1** (or **ad2**); while the most interesting experiment is to include a non-adenine requiring mutant, such as one that requires tryptophan for growth. The inoculations can be done with an inoculation loop, sterilizing the loop afresh when sampling a different yeast. Use a *very small* (less than pinhead size) amount of yeast so as to avoid a false positive for growth on YMM. It is sensible to inoculate the plates in the order YMM, YMM+adenine, YCM, to avoid transferring small amounts of adenine to non-adenine containing plates. Alternatively, the inoculation can be done by using a small drop of a saline suspension of each yeast prepared beforehand in a bijou bottle. After inoculation, plates should be

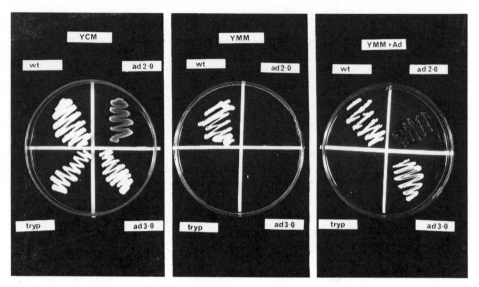

Fig. 4.7 Growth characteristics of wild type (wt), **ad2.0**, **ad3.0** and a tryptophan requir-ing mutant (**tryp**) of yeast, on YCM, YMM and YMM+adenine plates. Note that none of the mutants grows on YMM and that the tryptophan requiring mutant also does not grow on YMM+adenine (it *would*, however, grow on YMM supplemented with tryptophan).

left for 5 or so days in the laboratory at room temperature, or in an incubator set at 25°C, for adequate growth to occur. The wild type will be found growing on all three plates, while the adenine mutants will only be growing on YMM+adenine. A tryptophan requiring mutant would be found growing only on YCM (Fig. 4.7). Moreover, the **ad1** and **ad2** yeast will be red while all others will be white.

From this exercise students should come to realize that not only the colour but also the growth characteristics of the mutant yeast (e.g. no growth on YMM) are aspects of their mutant phenotype; and should become thoroughly familiar with the genetic basis of these phenotypic traits (Figs 4.3 and 4.4).

Yeast cells can be observed under the light microscope, by mixing a small amount of yeast into some cotton blue stain on a microscope slide before apply-ing a coverslip; or by putting a drop of yeast suspension directly onto a slide and examining the unstained and still living cells. In either case the cells should be studied with the x40 objective lens of the microscope. For unstained cells, close the iris diaphragm almost completely to increase contrast. Examine cells for their various phenotypes, especially cell size, shape and budding (Fig. 4.8). Note that adenine mutants cannot be distinguished from wild type cells on the basis of colour or any other feature when looking at individual cells.

2 Complementation Tests

Complementation is the process whereby two mutations in different genes, when combined in a diploid, supply one another's deficiency to produce a wild type phenotype. If we consider the general case for a diploid organism in which two individuals, each homozygous for a mutation in a different gene, are crossed,

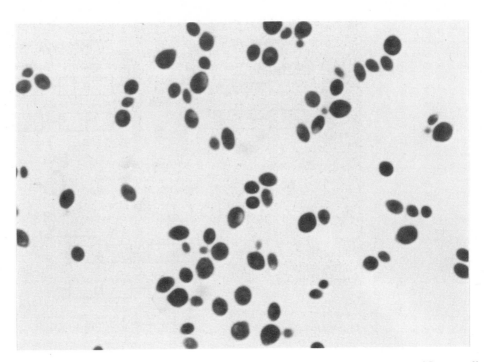

Fig. 4.8 Yeast cells as seen under the microscope after staining in cotton blue. Note small and large cells, round and oval cells, and that some cells were in the process of asexual reproduction by 'budding'.

then the F_1 will be heterozygous at both loci, as shown in Fig. 4.9a. In this situation, one parent has an abnormal phenotype due to a mutation in gene 1 while for the other parent the mutation is in gene 2. But the F_1 heterozygote has a normal allele (indicated as +) at both loci and displays the wild type phenotype. In terms of gene action we can easily understand that the F_1 codes for two normal enzymes from its normal alleles, so as to produce a normal phenotype, assuming, as is often the case, that the amount of enzyme coded by a single dose of each gene is sufficient for its normal catalytic function (i.e. the enzyme is *haplosufficient*). Thus complementation takes place when a heterozygote is made between parents that are mutant in different genes (i.e. the genes are non-allelic). The different genes may or may not affect the same characteristic.

When two individuals are homozygous for mutations within the *same* gene (i.e. allelic mutants), however, complementation will not usually take place and the F_1 will have a mutant phenotype, as illustrated in Fig. 4.9b. In this case we are assuming that the two mutations have arisen independently and that they represent DNA changes *at different places* within the same gene (hence **a** and **a*** in gene 1 of Fig. 4.9b). We therefore need to represent both these sites when writing down the genotypes. We also see that the F_1 cannot produce a normal enzyme from the locus concerned because neither allele is normal. Allelic mutants will normally have similar phenotypes, because they affect the same gene.

The rule then is that mutations within the same gene fail to complement one another, whereas mutations in different genes do complement and give wild type

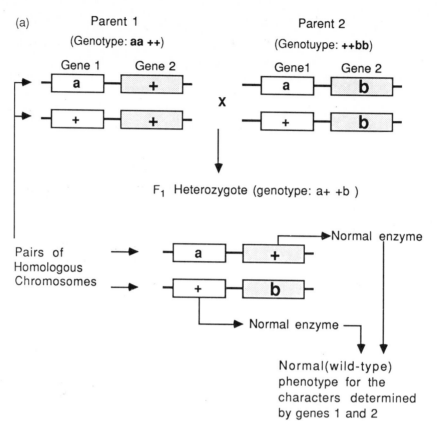

Fig. 4.9 Diagrammatic illustration of the biochemical explanation of (a) the phenomenon of complementation between non-allelic genes, and (b) the absence of complementation between allelic genes. See text for explanation.

phenotypes. Complementation tests are important in practical genetics, because quite often we isolate mutants which look similar (for example, both confer bright red eye colour to Drosophila flies), and we want to know whether the mutants belong to the same gene or to different genes. By crossing the mutants to make an F_1 we can obtain the answer. Non-allelic mutants will complement; allelic ones usually will not.

In yeast, complementation tests are made simply by mixing together two haploid strains of opposite mating types: they will readily fuse together in pairs to form diploid cells, which then multiply to form diploid colonies. In the following, we first describe a simple exercise that demonstrates complementation. The exercise will encourage students to understand the biochemical–genetic basis of the phenomenon and of what dominance means (in this case) in biochemical–genetic terms. Then we describe an experiment in which a number of 'unknown' red adenine requiring mutants are tested for whether they are mutant in the same gene or in different genes.

A simple exercise with cultures of known type
There are two steps to this demonstration of complementation.

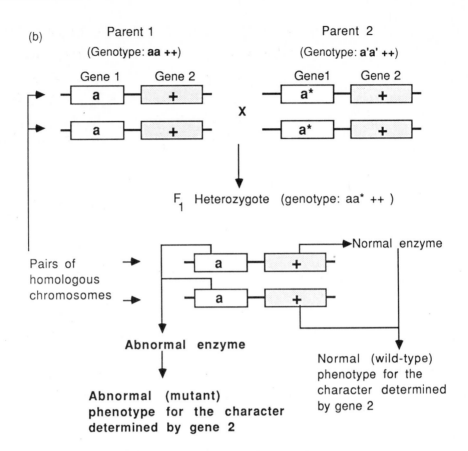

1 In this first step, each student is given a YCM plate and has access to YCM cultures of *opposite mating types* of **ad1** and **ad2** strains (*any* **ad2** mutant will suffice). The underside of the YCM plate is marked with a large capital Y, one fork of which is labelled **ad1**, the other **ad2**. Then using a sterile loop the **ad1** (or **ad2**) yeast is first streaked over the appropriate part of the agar, including down along the stem of the Y. After re-sterilizing the loop (!) the other yeast is streaked along its appropriate fork and then down along the stem of the Y so as to mix the two mutant strains. Avoid back-tracking along the forks once the two strains have been mixed in the stem. The cultures are then left for a few days, after which red growths of unmated haploids will be found along the forks of the Y (as controls) and white diploid growths (plus some red colonies from unmated haploids) along the stem of the Y (Fig. 4.10).

2 The second step in the exercise is to test the growth characteristics of the white colonies. This is done by making a suspension of white diploid cells in sterile saline, and also one from wild type yeast (to act as a control). YCM and YMM plates are marked on the back with a marker pen and the halves labelled 'wild type control' and 'diploid test'. Then each half is inoculated appropriately by spreading on it one drop of suspension, using a sterile loop. After 5 days, growth of the wild type *and* of the diploids will be observed on both culture plates.

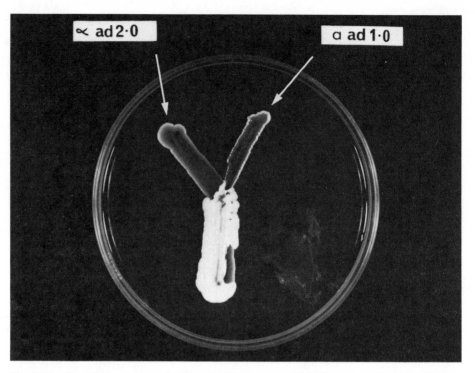

Fig. 4.10   Complementation of **ad2** and **ad1** yeast on YCM. The two forks of the 'Y' are red (with young pink growth around the edges), while the stem of the 'Y' is mostly composed of white complementing diploids, which formed from matings between the two adenine requiring mutants.

Both the colour and the growth characteristics of the diploids then should be explained in biochemical–genetic terms.

An experiment with 'unknowns'

A technically more demanding and interesting experiment makes use of four independently isolated red-coloured ad^- mutants having the same phenotype (colour and growth characteristics) and we attempt to establish whether they are all mutant in the same gene or in different genes.

To carry out this experiment each student or group of students is provided with the following:

Cultures of four red ad^- mutants growing on YCM plates. The strains to use are **ad1.0, 2.0, 2.1** and **2.6**; though when given to students they are best labelled 'unknowns 1 – 4'. *Both* mating strains of each mutant must be made available (eight cultures in all).

Eight sterile bijou bottles each about half filled with sterile saline (or a supply of sterile saline for students to add to the bijou bottles).

Plates of YCM which have been prepared several days in advance and allowed to become partially dry.

The procedure is as follows.

(a)

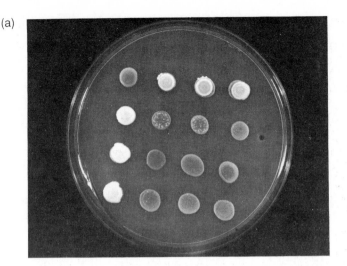

(b)

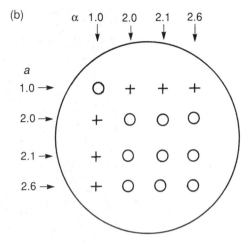

Fig. 4.11 (a) A complementation test culture plate showing the 16 diploid colonies produced after inoculation with all pairwise combinations of four ad⁻ haploid yeast mutants. (b) Results of the above complementation test, recorded as '+' for complementing (white growth) and '0' for non-complementing (red growth). The diagram is also annotated to show the origin of each diploid colony from pairwise matings of haploid strains.

1   Make a concentrated saline suspension of each culture, taking care to avoid cross contamination between strains and mating types. Label the suspensions carefully.

2   Mark the backs of the YCM plates at 16 places, labelling each mark according to the scheme shown in Fig. 4.11b. If the cultures are to be handled as unknowns, label the marks accordingly.

3   Shake the suspension of **ad1** (unknown 1) yeast mating type *a* to re-suspend the cells. Using either a sterile pipette or loop place a *small* (half) drop of suspension onto the surface of the agar on each of the four top marks (refer again to Fig. 4.11). With a fresh or re-sterilized pipette or loop repeat the process in the appropriate positions for **ad2.0**, then **2.1** and then **2.6**, by which time all 16 marks should be covered.

   Allow several minutes for the drops to dry and then cross-inoculate *on top of the first drops* using the α mating type of each strain, so as to produce all possible pairwise matings of the four strains. Allow 5 days for growth to occur.

   When the strains are combined in this way the haploid cells fuse to give diploid cells, which grow into discrete circular colonies. Complementing pairs of mutants will produce white (diploid) colonies while non-complementing pairs will produce red colonies (Fig. 4.11a). The few white spots that may be found in some of the non-complementing red colonies will be due to back mutations having occurred in the stock cultures while they were in storage (or due to some cross-contamination of the strains during suspension making or inoculation of the plates!). When the cultures have grown sufficiently they may be stored in a refrigerator until it is convenient to examine them. The results may be recorded using a '+' for the complementing pairs and a '0' for the non-complementing ones (Fig. 4.11b). Note that the plate carries a double series of all pairwise combinations of crosses, separated by the diagonal line shown in Fig. 4.11b.

The conclusions to draw from the result shown in Fig. 4.11a are that although all four strains have identical phenotypes, one strain (**ad1**) must have a mutation in a different gene from the other three, and that these three are all allelic mutations in the same gene (**ad2**).

3 Induction of Mutations using Ultraviolet Light

Gene mutations arise spontaneously through changes in the nucleotide sequence of DNA comprising a gene, often as a result of errors in DNA replication. Mutations may also be induced at high frequencies when the genetic material is treated artificially with certain chemicals or radiations. Ultraviolet light (UV) is a mutagen. The UV rays are absorbed by DNA, causing adjacent thymine molecules to become linked together directly to make a bulge in the normally regular DNA double helix. The bulge is thought to disrupt the bonding of these thymine bases with their complementary adenine bases in the opposite strand of the DNA. In daylight, DNA repair enzymes cut out these dimers and replace them with normal thymine bases. Dimers that are not replaced can create problems during DNA replication, leading subsequently to abnormal nucleotide sequences, i.e. to a mutation.

In yeast, the red-coloured ad⁻ mutants can be used for the visual detection of additional mutations induced by UV light. UV treatment of **ad1** cells will give *white* mutants which are of three main types:

1 *Double mutant* in which a single new mutation has been induced by UV treatment in another gene at an earlier step in the pathway of adenine synthesis. Such a double mutant will still be adenine requiring, but since the block is at an earlier step no red pigment is formed and the phenotype is therefore white (Fig. 4.12).
2 *Back mutant* (revertant) in which the mutation at **ad1** has reverted to its original wild type state as a result of UV treatment. These mutants no longer require adenine, and they are white since the red pigment no longer accumulates with the release of the block to E.
3 *Triple mutant* in which *two* new mutations have been produced by UV treatment, one at an earlier step in the adenine pathway, as in **1** above, the other in a gene involving some other metabolic pathway.

Amongst the red colonies that grow after UV treatment, some will also be double mutants, but with the second mutation *not* involving the adenine pathway. These mutants will have a requirement for a different substance, such as histidine, as well as for adenine. Since the block at E remains this class of double mutant will remain red and will be impossible to detect visually.

The following exercise may be used simply as a *demonstration* of the effects of UV treatment of yeast cells, especially its ability to induce mutations. Here students can be provided with all the background information given above, from

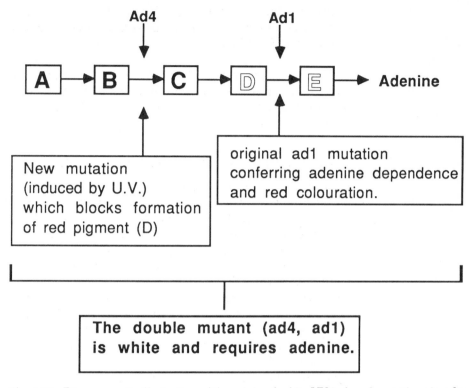

Fig. 4.12 Diagrammatic illustration of the origin of white UV induced mutations in **ad1** yeast. See text for explanation.

which they proceed to demonstrate the origin of white mutants by UV treatment of **ad1** yeast. For a more challenging exercise, students can be given just a minimum of background information (Fig. 4.4, for example, but not Fig. 4.12), from which they proceed to analyse and interpret a number of white mutant colonies they produce by UV treatment of **ad1** yeast.

The following equipment will be required:

YCM culture of **ad1** yeast,
50 ml sterile saline in a small flask or bottle,
Haemocytometer,
Supply of 9 ml sterile saline in universal or bijou bottles,
Sterile 1 ml pasteur pipettes,
YCM plates,
UV lamp of low output (about 30 W), safety glasses and disposable gloves.

The procedures to adopt are as follows. Remember to use sterile technique throughout; to shake suspensions before you sample from them; and to label suspensions and plates very carefully.

Producing the mutants
This is done by exposing yeast cells to UV light for 2–4 min, as follows.

Preparing the yeast suspension

Use about $1/4$ of a loopful of cells to make a concentrated suspension of **ad1** yeast in a 50 ml sterile saline. Label *'undiluted suspension'*. Estimate the number of cells per ml in this suspension, using a haemocytometer slide.

Make serial dilutions to obtain a final concentration of cells of about 2×10^5 per ml. In making these dilutions proceed by stepwise additions of 1 ml of suspension to 9 ml of fresh saline. For the final step a 1 : 9 dilution may not be appropriate and some calculation will then be necessary to get the desired final concentration. Label the suspension *'diluted suspension'* or simply *'10^5'* (as an abbreviation of 2×10^5).*

Setting up controls

These are used for monitoring the effect of UV treatment on cell viability and for measuring the spontaneous mutation frequency.

Using the dilute (10^5) suspension, make two further 1 : 9 dilutions labelling them *'control 1'* and *'control 2'* (or simply *'10^4'* and *'10^3'*). These two dilutions are designed to produce comparable numbers of colonies on the control and treatment plates after growth.

Using a sterile 1 ml pipette, inoculate a YCM plate with 0.1 ml of the diluted (10^5) suspension and carefully spread the inoculum over the entire surface of the agar with a sterile glass spreader. Repeat this procedure for each of the two control (10^4 and 10^3) suspensions, using a newly sterilized pipette at each stage. Label the plates *'diluted, 10^5 '*, *'control, 10^4 '* and *'control, 10^3 '*, respectively.

Irradiation

Turn the UV lamp on for a 5 min warm-up period. During this period, inoculate three or more YCM plates each with 0.1 ml of the *diluted (10^5)* suspension and spread the inoculum thoroughly over the agar.

Irradiate these plates for 2-4 min with the lids *off* (UV does not easily pass through plastic). Varying the exposure times over the 3 or more YCM plates is important. Label the plates 'UV treated, 2 min.', 'UV treated 3 min', etc. Replace the lids of the agar plates and incubate them for 7–10 days (some mutants grow more slowly than others).

Results

Examine control and treated plates (Fig. 4.13). Note the effect of UV on cell viability. Record the total number of colonies and the number of white colonies for both the control (10^4 and 10^3) and treated plates. Count any pale pink colonies as if they were red (though they make for interesting separate study). Estimate the spontaneous and induced mutation frequencies (= the number of mutant colonies divided by the total number of colonies). An induced mutation frequency of around 5% can be expected.

---

*Dilution is essential because the treatment times given below produce a 90% or more kill of cells and leave about 50 colonies when the survivors have grown on agar plates. If the concentration of cells is too high or too low it is difficult to isolate or find enough mutant cells.

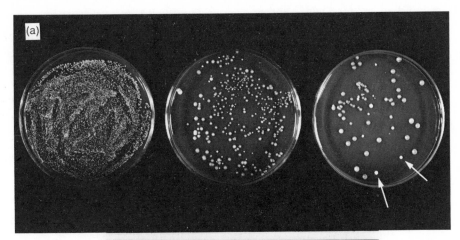

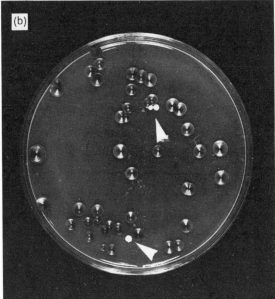

Fig. 4.13 (a) From left to right, undiluted, control 10^3 and UV treated cultures of **ad1** yeast on YCM. Note (i) the very high kill rate resulting from the UV treatment and (ii) two colonies (arrows) which are white and thus represent UV induced mutations. All other colonies are pink or red. (b) An older treatment plate, showing sharper size and colour differentiation of three white (mutant) colonies amongst the more numerous red colonies.

Classifying the mutants

Procedure

Circle and label the mutant colonies 1, 2, 3, etc. Test each mutant for adenine requirement by inoculation onto YCM (control), YMM and YMM+adenine. Up to eight colonies can be tested on the one plate (Fig. 4.14). Allow 3–5 days for growth.

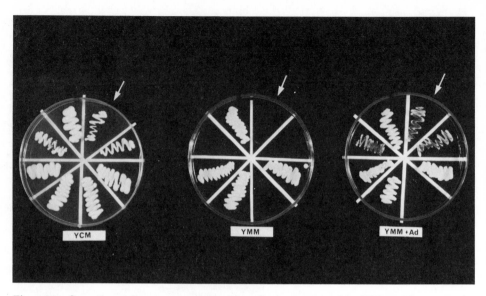

Fig. 4.14 Growth requirement test of 8 *white* colonies produced by UV irradiation of **ad1** yeast. The arrows locate mutant number 1 on each plate. Note that 4 mutants grew on YMM *and* YMM+adenine plates. Another 3 mutants did not grow on YMM but did grow on YMM+adenine, while another one grew neither on YMM nor on YMM+adenine. See text for interpretation.

Classify the mutants according to their growth characteristics, namely either
1   Growth on all three plates (**ad1** revertants), or
2   Growth on YCM and YMM+adenine but *not* on YMM (double mutants, requiring adenine), or
3   Growth on YCM but no growth on YMM nor on YMM+adenine (triple mutants, requiring adenine *and* some other substance).

Results and interpretation
Pool results over the whole class. Consider the relative numbers of the three types of mutants. Actual results are shown in Table 4.1.

Table 4.1   Numbers of three types of mutants obtained from 13 agar plates of **ad1** yeast irradiated with UV for 3 or 4 min.

Mutant type	Number of mutants
Revertant	1
Double	30
Triple	3
Total	34
Total number of colonies on treated plates	884

Since the double mutants each derive from only one new mutational event they are much more common than the triple mutants, each of which derives from *two* new mutational events. The revertants are rare, because each requires a change in one *particular* (**ad1**) gene.

The induced mutation frequency was approximately 4% (34/884).

REFERENCES

Fincham J.R.S., Day P.R. and Radford A. (1979). *Fungal Genetics*. California, University of California Press.

Hicks J. (Ed.) (1986). *Yeast Cell Biology*. California, A.R. Liss.

Spencer J.F.T., Spencer D.M. and Smith A.R.W. (Eds) (1983). *Yeast Genetics, Fundamental and Applied Aspects*. Berlin, Springer-Verlag.

Chapter 5 GENETICS WITH SORDARIA

The exercises in this chapter are suitable for all levels of study, from senior school to university undergraduate. Sordaria is a common saprophyte living on dead organic matter and is perfectly safe to use in laboratory classes. There are no particular safety problems with handling the material. The only precautions which need to be taken are to follow the rules on the safe handling of micro-organisms which are given in the Appendix.

INTRODUCTION AND BACKGROUND

Certain fungi have features which make them particularly suitable for genetic studies. Two that are most useful and easy to handle are *Sordaria fimicola* and *Sordaria brevicollis*. These two species are haploid throughout their life cycle, except for a brief diploid phase which precedes meiosis. Their other main attraction is that when they undergo sexual reproduction the products of a single meiosis *are held together* as a row of haploid ascospores within an elongated sac (the ascus). The ascus and its spores can easily be seen under a low power (x10) microscope objective. Mutants are available that have ascospore colours that are different from the normal wild type black ones. Each ascospore determines its own pigment synthesis. When crosses are made between two different spore colour strains a hybrid ascus will have ascospores of both colours; and the arrangements of spores in the asci show us how their spore colour genes segregate and recombine in a single meiosis.

Life Cycle of Sordaria

Sordaria is a saprophytic fungus that is easily grown on a medium of cornmeal agar in petri dishes. It grows as a typical mould made up of numerous branching hyphae that spread over the surface of the agar and make up the vegetative body of the fungus known as the mycelium. The mycelium is not strictly a cellular structure, but rather is composed of hyphae that are multinucleate and divided up by septae through which the nuclei can migrate. The nuclei in the mycelium are haploid. A small piece of mycelium inoculated onto the centre of a petri dish grows very quickly at room temperature and radiates out to cover the plate in 4–5 days. The two species that are commonly used in practical genetics are *Sordaria brevicollis* and *Sordaria fimicola*.

The species *S. brevicollis* is *heterothallic*, in that for sexual reproduction to occur strains of opposite mating type have to be grown together. The life cycle of *S. brevicollis* is shown in Fig. 5.1. Cultures started off from small pieces of mycelium will complete the life cycle and produce mature fruiting bodies in about 10 days at 25°C. When two mating types are inoculated onto the same dish hyphae of

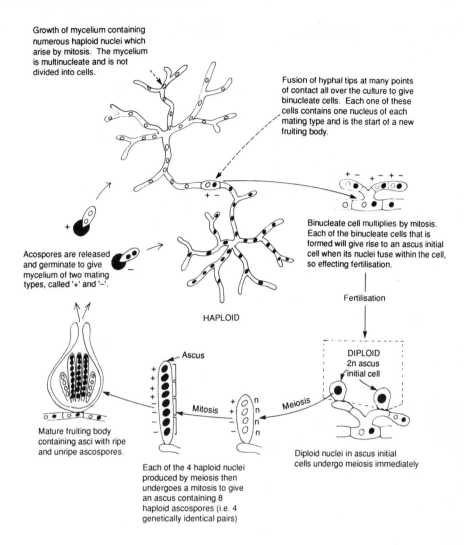

Growth of mycelium containing numerous haploid nuclei which arise by mitosis. The mycelium is multinucleate and is not divided into cells.

Fusion of hyphal tips at many points of contact all over the culture to give binucleate cells. Each one of these cells contains one nucleus of each mating type and is the start of a new fruiting body.

Binucleate cell multiplies by mitosis. Each of the binucleate cells that is formed will give rise to an ascus initial cell when its nuclei fuse within the cell, so effecting fertilisation.

Acospores are released and germinate to give mycelium of two mating types, called '+' and '−'.

Fertilisation

HAPLOID

DIPLOID
2n ascus initial cell

Ascus

Mitosis

Meiosis

Mature fruiting body containing asci with ripe and unripe ascospores.

Diploid nuclei in ascus initial cells undergo meiosis immediately

Each of the 4 haploid nuclei produced by meiosis then undergoes a mitosis to give an ascus containing 8 haploid ascospores (i.e. 4 genetically identical pairs)

Fig. 5.1 Simplified diagram of the life cycle of the fungus *S. brevicollis*. The organism is haploid for most of its life cycle. The diploid phase is restricted to the ascus initial cells, which undergo meiosis almost immediately after they are formed at fertilization. The gametes are produced by mitosis and are in the form of pairs of nuclei which occur in binucleate cells in developing fruiting bodies.

opposite mating types will fuse at many places all over the culture. Once the hyphae have fused, by breakdown of adjoining walls, they form a proper cell which encloses a pair of nuclei, one from each mating type. Each binucleate cell then multiplies several times, by 'conjugate mitosis', to give a group of several such cells. Fertilization then takes place by the fusion of pairs of nuclei, to form diploid ascus initial cells (Fig. 5.1). These ascus initial cells are the only diploid phase in the life cycle, and they undergo meiosis almost immediately to give four haploid nuclei within each developing ascus. Following meiosis each of the four haploid nuclei goes through a mitosis, and this produces in each mature ascus eight haploid ascospores in four pairs, members of each of which are genetically identical. A fruiting body (perithecium) forms around the group of ascus initial cells during their development, and each mature perithecium comes to contain numerous asci all originating usually from a single binucleate cell. The perithecium is a dark, flask-shaped structure about the size of a pin head. Eventually the ripe ascospores are discharged from the asci and dispersed.

The life cycle of *S. fimicola* is the same as that of *S. brevicollis* except that this species is *homothallic*. There are no mating types and a culture started from a single spore will undergo hyphal fusions and form binucleate cells containing genetically identical pairs of nuclei. These will then develop into fruiting bodies. When two different spore colour mutants are grown together in the same culture some hyphal fusions will give rise to cells with identical pairs of nuclei, and some to cells with one nucleus from each strain, so that both 'selfing' and 'crossing' take place in the same culture.

Ascus Formation

In both species the feature of sexual reproduction which is of special interest to us is the fact that each mature ascus contains the paired cell products from *one meiosis*, laid out in order in relation to the first and second meiotic divisions (Fig. 5.2). Since these ascospores have character differences in relation to pigmentation of their cell walls, they provide a unique opportunity for investigations on how one or more pairs of alleles segregate and recombine during a single meiosis. To make these investigations we have to 'cross' strains with different coloured ascospores, so that the diploid nuclei in the ascus initial cells are heterozygous at the loci concerned.

PROCEDURES

Growing Sordaria

Sordaria is grown on cornmeal agar under sterile conditions. Stock cultures are

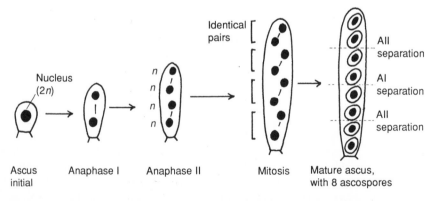

Fig. 5.2   Diagram showing how the arrangement of ascospores in a mature ascus of Sordaria reflects chromosome separation in anaphase I and anaphase II of meiosis. The plane of division of anaphase I is between the four spores in the top half and the four spores in the bottom half of the ascus. The anaphase II plane of division is between the two pairs in each of the halves.

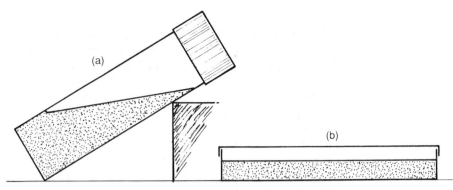

Fig. 5.3 The culture medium for growing Sordaria is contained in specimen tubes for maintaining stock cultures (a), and in petri plates for making crosses between strains (b). To make agar slants (a) the specimen tubes are placed at an angle while the agar cools and solidifies.

maintained on agar slants in universal screw-cap specimen tubes, whereas in crossing experiments petri dishes are used.

To prepare the culture medium, 17 g of cornmeal agar + 1 g of yeast extract is added to 1 litre of water and boiled up in a saucepan. Once the ingredients have dissolved the solution is transferred to bottles (with screw-caps loosened) and sterilized at 15 p.s.i. pressure for 15 min in an autoclave or domestic pressure cooker. When the medium has cooled sufficiently to handle it is poured into petri plates or stock tubes and left to solidify. Stock tubes are poured about one-third full and then placed at an angle so that the agar sets at a slant (Fig. 5.3). The agar can be poured in the laboratory provided it is done quickly on a clean bench and in the absence of draughts. A few extra tubes should be made to allow for wastage by contamination. No special facilities are required. Glass petri plates and screw-cap tubes must also be sterilized before use. Sterile disposable containers can also be used. A litre of medium makes about 30 thickly poured plates, and these will keep for 2 weeks or so in a refrigerator if desired.

Strains of Sordaria are normally received in sealed specimen tubes. The surface of the agar will be covered by mycelium, and fruiting bodies may also be present. When used the cultures are opened, and a small block of agar is removed with a scalpel (after sterilizing the blade in a bunsen flame) and transferred to the centre of a fresh plate of agar. The mycelium will grow quickly at room temperature (or in an incubator at 25°C) and cover the agar in 4–5 days.

Maintaining Stock Cultures

Sordaria fimicola

In *S. fimicola* stock cultures of the grey (g) strain and its corresponding wild type strain (g^+) are grown in separate specimen tubes on agar slants. The mycelium is allowed to grow for about 1 week with the screw caps loosened until perithecia begin to form, and then the caps are screwed up tightly and the tubes kept in a refrigerator at 2–4°C. Tubes can be kept in this condition for at least 6 months, after which the cultures should be grown on petri plates to revive them and then put back into fresh tubes for another 6 months. In this way the cultures can be kept indefinitely (Fig. 5.4).

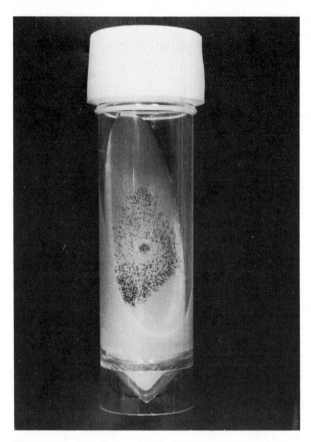

Fig. 5.4 Stock culture of *S. fimicola* growing on an agar slant in a specimen tube. The culture has been allowed to grow for 10 days at 25°C (with the screw cap loosened) in order to form fruiting bodies and mature ascospores. After the screw-cap has been tightened the culture will remain viable for 6 months or more in a refrigerator at 2–4°C.

Sordaria brevicollis

In theory stock cultures of *S. brevicollis* should be able to be maintained by regular sub-culturing of the vegetative mycelium of each mating type strain, the cultures being kept in the same way as those of *S. fimicola*.

In practice, however, this does not work very well. The main problem is that the single mating type strains kept in this way gradually lose their capacity to mate with strains of the opposite mating type, i.e. they become infertile. To avoid this it is necessary to take them through a sexual cycle from time to time and then to re-isolate the mating types by growing them up from single ascospores. To do this for the wild type black-spored strain, for example, the '+' and '-' mating types are grown together to give perithecia containing all black ascospores. Since mating type is controlled by a single gene the two alleles will segregate out to give four '+' and four '-' mating type ascospores in each ascus. These ascospores can be spread out thinly on agar plates and allowed to germinate. The mycelium which grows from single spores is isolated into their individual specimen tubes. These single spore cultures then have to be tested one against another to classify their mating type. The same procedure is adopted for the buff and yellow-spored mutants.

An alternative and easier method of maintaining sexually viable cultures is to keep the two mating types of each strain *together*, so that they produce fruiting bodies and sexual spores in the stock cultures. When the strains are grown up ready for crossing the ascospores are sub-cultured as well as the mycelium, and since both mating types are present perithecia and asci will be produced. Crosses between strains are then made by inoculating ripe ascospores as well as mycelium onto the crossing plate, where both mating types of both strains will be present. The crossing plate will then produce a mixture of selfed perithecia of both parental types, as well as hybrids. The hybrid ones will then have to be searched for.

EXERCISES

1 With *Sordaria fimicola*

Two strains of this species are widely used in teaching genetics. One strain has normal black ascospores and the other one is a mutant with ascospores that are virtually colourless (described as grey). The organism is haploid so the genotypes are written as g^+, for black spores, and g for grey spores (see Chapter 3 for further information on gene symbols). When the strains are grown singly the g^+ genotype produces perithecia containing asci with only black ascospores, and the mutant (g) gives perithecia with only grey ascospores.

Making crosses
Crosses are made by inoculating mycelium of the two strains together on the same petri dish (Fig. 5.5). After 7–10 days or so (depending on temperature) fruiting bodies will appear all over the culture, but the ones that result from crossing, having both black and grey spores, will only be found across the middle of the plate where mycelia of the two strains have grown together. The others will be *selfed* perithecia, containing either all black or all grey ascospores. The only way to tell whether or not a perithecium is of hybrid origin is to examine its contents under the microscope.

Examination of asci
Mature cultures will have perithecia that contain ripe asci, as well as some immature perithecia that are still in the process of development. To examine the asci, pick off four or five large perithecia with a needle and place them on a microscope slide in a drop of 45% acetic acid. It is best to take perithecia from the periphery of the culture where the two strains have grown together, as these are the ones most likely to be of hybrid origin.

Place a coverslip over the perithecia on the slide and then press down with the point of the needle directly over each perithecium in turn. In this way the fruiting body bursts and the asci are squeezed out, rather like the spokes of a wheel (Fig. 5.6). Finding hybrid perithecia requires some patience and thus sometimes the preparation of several slides. Resist the temptation to put a large number of perithecia onto the same slide, for this will make it impossible to squash them out properly.

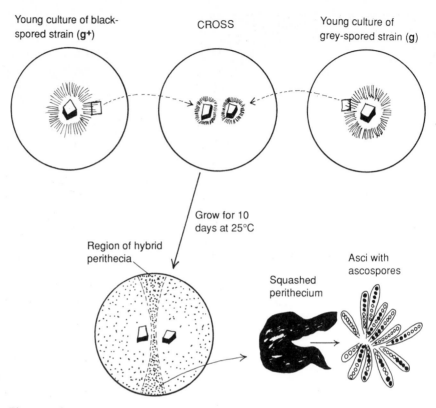

Fig. 5.5 Crossing black-spored and grey-spored strains of *S. fimicola*. The two parent strains are grown as separate cultures for 4–5 days in order to produce an abundance of actively growing mycelia. Pieces of mycelia are then transferred via small blocks of agar to the centre of the crossing plate. After 10 days at 25°C the crossing plate will be covered with mycelia and fruiting bodies. Hybrid perithecia will be formed across the centre of the plate where the mycelia of the two parental strains have grown together.

Apart from finding hybrid perithecia it is, of course, necessary to produce cultures at the right stage of development on the right day. If cultures are too young then no ascospores will be present, or else they may be immature and all colourless. If they are too old the asci will have released their spores (producing a black 'cloud' on the inside of the lid). It is necessary, therefore, to monitor the cultures part way through the second week by squashing out a few sample perithecia. To be on the safe side it is best to have them maturing a few days in advance of the class and then to 'hold' them at the right stage in a refrigerator (2–4°C). Another useful tip is to start off two sets of crosses with a couple of days in between, and so stagger their development.

Good slides should have their coverslips surrounded with nail varnish or rubber solution to prevent drying out. Fig. 5.6 shows what we can expect to see from squashing hybrid perithecia.

Observations

Individual students should prepare squashes of hybrid asci and study their six arrangements of black and grey ascospores. These arrangements in their typical

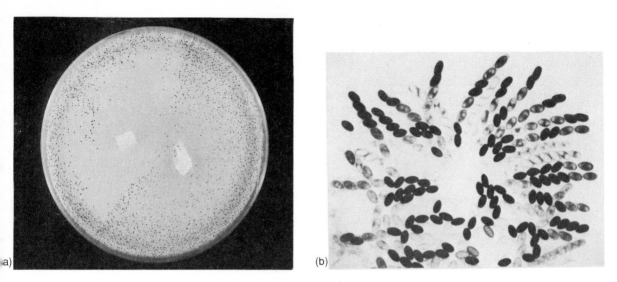

(a)   (b)

Fig. 5.6 Photographs of *S. fimicola*. (a) 10-day-old culture from the cross black x grey, showing perithecia. (b) Asci from a squashed hybrid perithecium, showing arrangements of ascopores that result from segregation of the **g⁺** and **g** alleles during meiosis.

relative frequencies, taken from a set of class data, are shown in Fig. 5.7.

Each student should record the arrangements in a sample of about 50 asci (for a class of ten), being as objective as possible in taking the sample. Distinguishing the top and bottom of the ascus, even if the asci have become disorganized during squashing, can be done by reference to the truncated shape of the top of the ascus. However, not a great deal of information is lost if the six spore arrangements of Fig. 5.7 are simply recorded as belonging to two groups only, namely either the 4 : 4 type, in which four ascospores of the same colour are present in each half of the ascus, or the 2 : 2 : 2 : 2 type, which have both colours of ascospores

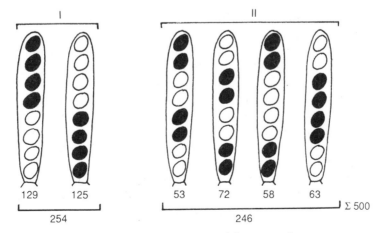

Fig. 5.7 Diagrammatic representation of the types of ascospore arrangements found in *S. fimicola* from the cross of black-spored (**g⁺**) x grey-spored (**g**) strains. Numbers below asci refer to typical frequencies and are taken from a set of class data. See text for further explanation.

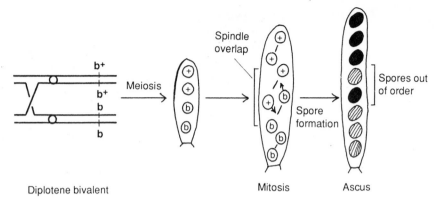

Fig. 5.8 Diagram showing how spindle overlap at mitosis can lead to asci in which the ascospores are out of order.

present in each half of the ascus (see below for further explanation).

It is sensible to direct student attention to the following issues prior to their analysis, to avoid undue confusion.

1 Some asci of a perithecium will have all colourless ascospores, because the asci within one fruiting body do not all develop at the same time. Moreover, some ascospores will be dark green in colour, this being an intermediate stage in the synthesis of the black pigment.

2 The occasional perithecium may have developed from more than one binucleate cell, and this can give rise to some asci having all black or all grey ascospores alongside the hybrid ones.

3 Occasionally some unusual asci will be found. One such unusual ascus has four black and four grey spores but these are out of order, due to spindle overlap. What happens is that during the mitosis which follows meiosis the spindles may overlap one another and the division products end up out of sequence. An example is shown in Fig. 5.8. These irregularities can usually be sorted out quite easily, or else such asci can be by-passed during scoring. (The problem is more prevalent in *S. brevicollis* than in *S. fimicola*.)

4 A few unusual asci will be found that do not contain the four black and four grey spores expected by strict Mendelian segregation. There may be six black : two grey, or vice versa, in various arrangements; or five of one colour and three of the other. These aberrant asci arise from a process called *gene conversion*, which occurs when recombination takes place within the **g** locus itself. The reader is referred to a more advanced text for an explanation of this phenomenon, which is best left for study in advanced genetics classes.

Interpretation and analysis

The principle of segregation (Mendel's First Law)

The first point to note is that these hybrid asci provide a direct and convincing demonstration of the *principle of segregation*, i.e. Mendel's First Law. Each heterozygous diploid nucleus yields four black and four white spores in each hybrid ascus (notwithstanding the rare exceptions noted in **4** above). This implies that

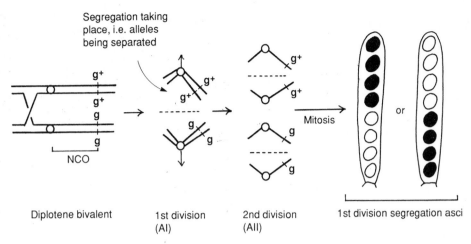

Segregation taking place, i.e. alleles being separated

NCO

Diplotene bivalent   1st division (AI)   2nd division (AII)   1st division segregation asci

Fig. 5.9 Interpretative diagram of the meiotic event which leads to first division segregation in a hybrid between black-spored (g^+) and grey-spored (g) strains of *S. fimicola*. (NCO = no crossing over between the gene locus and the centromere.)

the pair of spore colour alleles have separated from each other during meiosis so that now only one is present in each cell (spore). To deduce this in diploid organisms such as Drosophila we have to look at the phenotypes of a large sample of F_2 or testcross progeny. Here we look directly at the haploid segregating products of an *individual meiosis*.

First and second division segregation
The arrangements of the ascospores within hybrid asci allow us not only to observe segregation but also to discover in which of the two divisions of meiosis segregation takes place. Consider first of all the 4 : 4 type asci. These are the result of *first division segregation*. This comes about when there is crossing over (chiasma formation) somewhere other than in the region of the bivalent between the gene locus and the centromere (Fig. 5.9).

The two arrangements shown in Fig. 5.9 are basically identical: the difference between them depends simply upon which way round the bivalent attaches to the spindle at the first division of meiosis. Indeed, if a large number of the two types of 4 : 4 segregations are scored in an unbiased way it will be found that they occur in equal frequencies, implying that bivalent attachment to the spindle is random.

The 2 : 2 : 2 : 2 type asci are the result of *second division segregation*, which occurs when crossing over *does* take place in the region of the bivalent between the gene locus and the centromere (Fig. 5.10). The four patterns shown in Fig. 5.10 are basically identical. They occur with equal probability and result from the random way in which the bivalent (at the first division of meiosis), or the half-bivalents (at the second division), attach to the spindle with respect to the top and bottom of the ascus.

We see, therefore, that segregation of alleles takes place in *both* divisions of meiosis depending on whether or not chiasma formation takes place between the spore colour gene and the centromere.

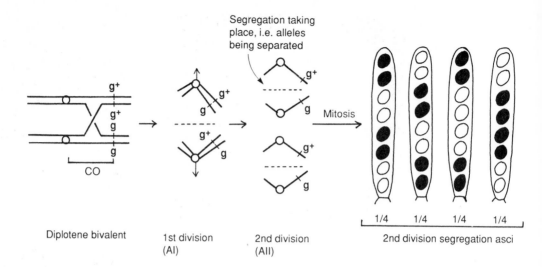

Fig. 5.10 Interpretative diagram of the meiotic event which leads to second division segregation in a hybrid between black-spored (g^+) and grey-spored (g) strains of *S. fimicola*. (CO = crossing over between the gene locus and the centromere.)

Mapping the centromere and the spore-colour gene
The set of class data shown in Fig. 5.7 is typical of that found for a sample of asci from a cross between the g^+ and g strains of *S. fimicola*. The percentages of the first and second division segregation asci are 51 and 49 respectively. This tells us that crossing over took place between the gene locus and the centromere in 49% of the meioses that were sampled; and that in the other 51% of meioses it took place in some other region of the bivalent. These values are characteristic for the g locus, because the locus occupies a fixed position in the chromosome, and because there is a certain probability that crossing over will take place in the region between the locus and the centromere. Other loci that are closer to or further away from the centromere will have correspondingly lower and higher second division segregation frequencies.

We can use this information on the percentage of second division segregation to map the position of the centromere relative to the g locus. This is done on the usual basis that 1% of recombination = 1 map unit. To convert second division segregation frequencies to recombination frequencies we have to divide the second division segregation frequencies by 2. This makes the map distance comparable to those obtained by conventional genetic methods.

$$\text{Recombination \%} \atop \text{(locus–centromere)} \quad = \quad \frac{\text{\% 2nd division segregation}}{2}$$

Fig. 5.11 Linkage map of *S. fimicola* showing the relative positions of the centromere and the g locus, based on the proportion of asci that show second division segregation of g and g^+.

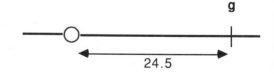

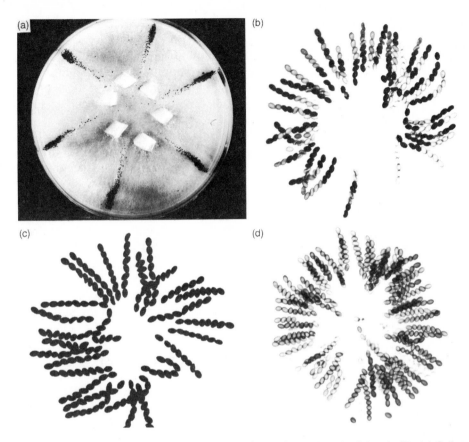

Fig. 5.12 Photographs of perithecia and ascospore formations in *S. brevicollis*. (a) Culture plate with perithecia. The plate was inoculated at the centre with mycelium of '+' and '-' mating types. Sexual reproduction has taken place along the radii of the culture where mycelia of opposite mating types have grown together. (b) Squashed out hybrid perithecium from the cross black x buff. All of the different spore arrangements referred to in the text can be seen. In a couple of asci the spores are out of order due to spindle overlap at the mitosis which follows meiosis. (c–d) Asci from cultures of the two parents used to make the cross shown in (b). The black-spored strain is shown in (c) and the buff strain in (d). The variations in colour intensity in (d) are due to differences in the stage of maturity of asci.

The reason for the division by two is that every time we have a second division segregation taking place only two of the four chromatids are actually recombined, because crossing over only involves two of the four chromatids.

In the example we are using here recombination between the **g** locus and the centromere = 49/2 = 24.5%. The linkage map is shown in Fig. 5.11.

2 With *Sordaria brevicollis*

Along with the wild type strain, *two* spore colour mutants are commonly used in this species of Sordaria, namely buff (**b**) and yellow (**y**). When crossed they give a spectacular demonstration of recombination by crossing over.

Making crosses

Three crosses are possible, namely (i) wild type (black) x buff, (ii) wild type x yellow and (iii) buff x yellow. The easiest way to make these crosses is to grow opposite mating types of two appropriate strains in the same culture. When this is done the only fruiting bodies that form are those of hybrid origin, where the mycelia of the two mating types have grown together (Fig. 5.12).

Observations and interpretations

In crosses involving a single gene the procedure for examining asci and interpreting the patterns of segregation of the two alleles is the same as that described for *S. fimicola*.

Black x buff (b^+ x b).
Hybrid perithecia from this cross will give asci containing four black and four buff ascospores. As before there will be two first division segregation arrangements and four arrangements resulting from second division segregation (Fig. 5.12). In this case the ratio of first : second division segregation (say 54 : 46) will be different from that in *S. fimicola*, because we are dealing with a different gene locus. The recombination frequency between the buff locus and the centromere is 23%.

$$\text{Recombination \%} = \frac{\text{\% 2nd division segregation}}{2} = \frac{46\%}{2} = 23\%$$
$$\text{(buff-centromere)}$$

The buff locus is therefore 23 map units from the centromere.

Black x yellow (y^+ x y)
The **y** locus is in the same chromosome as buff (**b**), but is located further from the centromere. In crosses between black x yellow we therefore find a greater frequency of second division segregations than we do for b^+ x b.

Buff x yellow (by^+ x b^+y)
The buff and yellow mutations are in two different genes. Since we are now con-

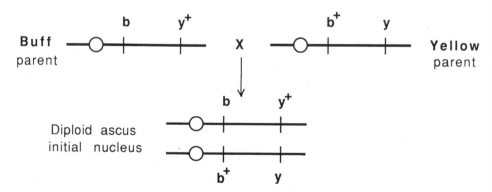

Fig. 5.13 Illustration of the *S. brevicollis* cross buff x yellow. Homologous chromosomes are represented as lines, in which are located centromeres (circles) and the buff and yellow gene loci, either as normal alleles (b^+ and y^+) or as mutant alleles (**b** and **y**).

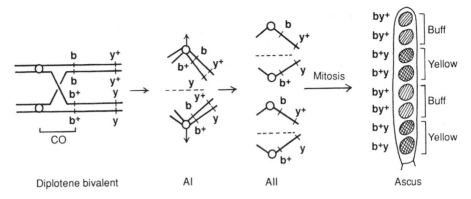

Fig. 5.14 Diagram showing the origin of a typical ascospore arrangement in a buff x yellow hybrid in *S. brevicollis*. The single crossover (CO) between the **b** locus and the centromere results in second division segregation for the alleles at both loci.

sidering two genes we have to write something down for both loci when we represent the genotypes. The buff strain is written as **by⁺**, since it carries the mutant allele at the **b** locus and has a normal allele (+) at the **y** locus. Similarly, the yellow strain is written as **b⁺y**, since it is normal at the **b** locus and mutant for **y**.

When a cross is made between buff x yellow strains (**by⁺** x **b⁺y**) the diploid ascus initial cells are heterozygous at both loci, as shown in Fig. 5.13. This figure also shows that the two genes are linked and are carried in the same arm of the chromosome concerned. The **y** locus is furthest away from the centromere.

When meiosis takes place in these heterozygous nuclei there is segregation for both pairs of alleles, and this leads to some interesting results. In all, 36 different arrangements of ascospores are possible, involving buff, yellow, black and white spores. We need concern ourselves with only a few of these, as follows.

The majority of hybrid asci will contain a mixture of buff and yellow ascospores only. The arrangements that are found will depend upon where crossing over takes place, and how the bivalents and half-bivalents are oriented on the spindles at the first and second divisions of meiosis. It should be quite easy to explain the origins of the various spore patterns on the basis of what was said earlier for the segregation at a single locus (Exercise 1). One example of an ascus with buff and yellow ascospores is shown in Fig. 5.14.

A minority of asci will contain black and white ascospores in addition to the buff and yellow ones. These are *recombinant* spores in that they arise when crossing over takes place between the **b** and **y** loci (Fig. 5.15). The black ascospores have the wild type alleles at both gene loci, which enables them to produce black pigment. The white ascospores are mutant at both loci, and this gives them a unique phenotype which can be visually distinguished from both of the single mutants. It is because the asci containing recombinant ascospores are in the minority that we know the two genes are linked. Had the genes been unlinked the two pairs of alleles would segregate independently and we would have as many black plus white ascospores as buff plus yellow one, which is not the case.

It is important to stress again that the value of these ascomycete fungi lies in us being able to observe the way in which pairs of alleles segregate and recombine

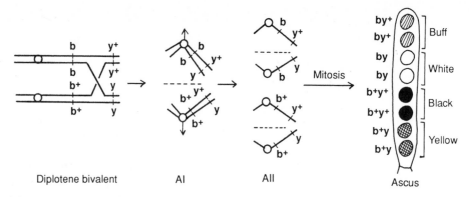

Fig. 5.15   Diagram showing the origin of recombinant ascospores in a buff x yellow hybrid in *S. brevicollis*. The single crossover occurred between the **b** and **y** loci.

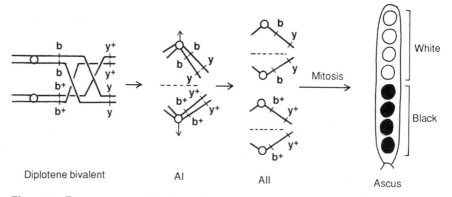

Fig. 5.16   Diagram to explain the origin of a rare type of ascus containing only recombinant ascospores from the cross of buff x yellow strains of *S. brevicollis*. Two crossover events occurred between the **b** and **y** loci, involving all four chromatids.

in a single meiosis. In the example shown above we can interpret the spore patterns only on the basis that crossing over takes place after chromosome duplication, and that it involves only two of the four chromatids. In other kinds of organisms (Drosophila, maize, mice, peas, etc.) it is not possible to demonstrate the properties of crossing over in this way, because we cannot trace the origins of any four F_2 or testcross progenies back to one meiosis.

A rare kind of ascus turns up periodically in a buff x yellow cross. It contains only recombinant ascospores, namely four black and four white! This is explained by the occurrence of a *double* crossover between the **b** and **y** loci, involving *all four* chromatids (Fig. 5.16).

REFERENCE

Fincham J.R.S. (1971). *Using Fungi to Study Genetic Recombination.* Oxford Biology Readers, eds J.J. Head and O.E. Lowenstein. Oxford, Oxford University Press.

Chapter 6   GENETICS WITH ASPERGILLUS

The exercises in this chapter are most appropriate for polytechnic and university first and second year levels: they may be too ambitious for most junior (A level or high school) classes in genetics. Students will find the exercises both enjoyable and instructive.

Aspergillus nidulans is a saprophyte that lives in soil and is not harmful to humans. However, it is advisable to keep culture plates sealed when culturing and studying them, especially so as to avoid inhaling clouds of dry asexual spores. After use, culture plates should be autoclaved before disposal. Follow the general procedures for the Safe Handling of Micro-organisms given in the Appendix.

INTRODUCTION AND BACKGROUND

Aspergillus nidulans is an Ascomycete fungus that grows as a saprophyte in soil. It is easy to grow in culture and has features which make it a very useful organism for studies in practical genetics. Aspergillus is haploid for most of its life cycle, it has both sexual and asexual modes of reproduction, the life cycle is short and there are many useful mutations affecting both the obvious phenotype (e.g. colour of conidia) as well as various biochemical features. The organism is safe to handle, and like yeast and Sordaria it provides training in the genetics and manipulation of micro-organisms.

Life Cycle

Asexual reproduction

The fungus is grown in the laboratory on sterile agar culture medium in petri dishes. Wild type strains can be grown on minimal medium (MM) and a few conidia (asexual spores) inoculated onto the centre of a plate will produce a colony that covers the whole dish in 5–6 days at 37°C. The colony is made up of an interconnecting network of hyphae (the mycelium) which spreads over the surface of the agar. The mycelium is divided into a number of compartments which contain numerous haploid nuclei. These nuclei multiply by mitosis as the colony grows. All over the surface of the culture aerial structures (called conidiophores) are produced which carry the asexual spores. In the development of a conidiophore finger-like projections (sterigmata) bud off long chains of uninucleate conidia. In the wild type strain these conidia have green pigment in their cell walls, thereby giving the colony its green colour. The chains of conidia grow from a single mother cell at the basal end of the chain; and each chain may contain up to 100 asexual spores. A whole conidiophore may have as many as 100 chains, and when mature the conidia blow away and may give rise to new fungal colonies (Fig. 6.1).

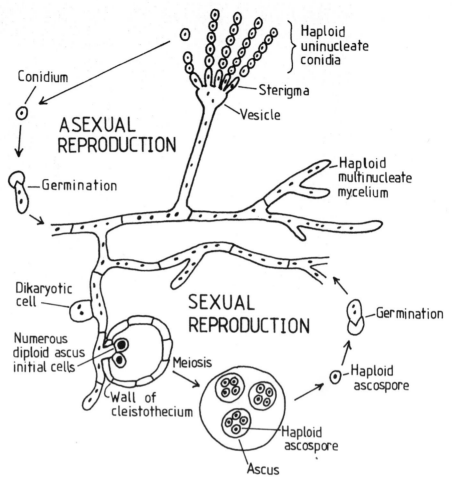

Fig. 6.1 Diagrammatic representation of the life cycle of *Aspergillus nidulans.*

Sexual reproduction

Aspergillus is naturally homothallic and sexual reproduction readily takes place under laboratory culture conditions. The sexual spores (ascospores) are found inside spherical black structures (cleistothecia) which form all over the surface of a culture plate after about 10 days incubation at 37°C. The details need not concern us here, but essentially what happens is that ascogenous hyphae are formed containing binucleate cells. Each clump of ascogenous hyphae originates from a single binucleate cell by conjugate mitoses; and in these binucleate cells nuclear fusion takes place to give rise to numerous diploid ascus initial cells formed in a developing fruiting body (the cleistothecium; Fig. 6.1). Each cleistothecium can have up to 100 000 ascus initial cells, all originating from one binucleate cell at the start of the sexual process. The ascus initial cells undergo meiosis, followed by mitosis, to give in each ascus eight haploid ascospores (four pairs, with members within each pair being genetically identical). The ascospores are minute and reddish-brown in colour, and when they are released and germinate the sexual process has completed (Fig. 6.1).

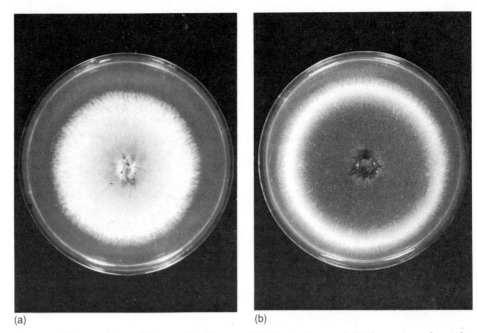

(a)   (b)

Fig. 6.2 Photographs of 5-day-old cultures of (a) the yellow and (b) the green-coloured parent strains of *Aspergillus nidulans* growing on complete medium.

PROCEDURES

Culturing Aspergillus

Conidia are normally used to start off a culture. Conidia are picked up on the tip of a sterile needle from a stock culture and inoculated onto the centre of a fresh agar plate containing a growth medium (see below). It is usual to hold the petri plate upside-down when making the inoculation to avoid showering spores too widely. The plates are then incubated upside-down to prevent condensation forming on the inside of the lid and then dropping onto the culture. Four to five days growth at 37°C is sufficient time for the surface of the agar to be covered completely with mycelium and asexual spores (Fig. 6.2). At least 10 days of growth are needed for the production of ripe ascospores. Once conidia and asco-spores have been formed the agar can be allowed to dry down completely, until paper thin. The ascospores will then remain viable for at least 1 year at room temperature. The standard way of keeping stock cultures is to grow the strains on agar slants of complete medium (CM) in universal tubes, and then to keep them in the tubes from one year to the next at room temperature. No complicated or time consuming methods are required. The ways in which the cultures are handled in experimental situations are described in the Exercises section.

Culture Media

Aspergillus Complete Medium (CM)

Sucrose	20.0 g
Difco 'Bacto' yeast extract	1.5 g

Difco 'Bacto' peptone	1.5 g
NaNO$_3$	6.0 g
Difco 'Bacto' casamino acids	1.0 g
Oxoid No. 3 Agar	15.0 g
Adenine	0.15 g
Vitamin Solution (see below)	10 ml
Water	1000 ml

For liquid nutrient omit the agar and sucrose.

Vitamin solution

Biotin	0.01 g
Pyridoxin HCl	0.01 g
Aneurin HCl (thiamin)	0.01 g
Riboflavin	0.01 g
p-aminobenzoic acid	0.01 g
Nicotinic acid	0.01 g
Water	100 ml

The solution should be kept in the dark, as riboflavin is photolabile.

Czapek's Minimal Agar (MM)

NaNO$_3$	2.0 g
KCl	0.5 g
MgSO$_4$.7H$_2$O	0.5 g
FeSO$_4$.7H$_2$O	0.01 g
ZnSO$_4$.7H$_2$O	0.01 g
K$_2$HPO$_4$	1.0 g
Sucrose	40.0 g
Oxoid No. 3 agar	15.0 g
Water	1000 ml

Supplements for nutritional mutants

| Adenine (ad) | 0.1 g/litre |
| Phenylalanine (phen) | 0.1 g/litre |

Water agar

| Agar | 20 g |
| Water | 1000 ml |

In all of the above media it is advisable to add two crystals of copper sulphate per litre to ensure full colouration of the asexual conidia.

EXERCISES

In the exercises to be described below two strains of Aspergillus involving three gene loci are used, namely

+ ad +   and
y + phen

The first strain has a wild type allele (+) for conidial pigmentation (giving green conidia), a mutant allele at the **ad** locus (giving a requirement for adenine) and the wild type allele (+) at the **phen** locus. This strain is therefore green in colour and requires either complete medium, or a minimal medium supplemented with adenine, in order to grow.

The second strain is mutant (**y**) at the locus for conidia pigmentation (making it yellow rather than green), and has the wild type allele (+) at the **ad** locus and a mutant allele at the **phen** locus. This strain is therefore yellow and requires either a complete medium, or a minimal medium supplemented with phenylalanine, in order to grow. The two strains are thus easily identified by colour.

1 Heterokaryons

When hyphae of Aspergillus meet they can fuse to produce interconnecting masses of mycelium. This makes it possible to combine nuclei of unlike genetic constitution from two (or more!) different parental strains in a common cytoplasm and thereby form a *heterokaryon*. Strains which carry only a single nuclear type are known as *homokaryons*.

Heterokaryons are made by mixing together the conidia of two parental strains in liquid CM (CM without the agar) overnight so that the spores germinate and produce a tangled mass of hyphae on the surface of the liquid, followed by plating out the mat of mycelium onto MM the following morning. On a minimal agar only the heterokaryons are able to grow.

Producing heterokaryons
The step-by-step procedure for producing heterokaryons is as follows.

1 Grow up the green (+ **ad** +) and yellow (**y** + **phen**) parental strains on CM for 4–5 days at 37°C to obtain cultures with masses of asexual conidia (Fig. 6.2).
2 Place liquid CM diluted to $1/4$ strength in four sterile bijou bottles filled to about two-thirds of their volume. Two bottles are for mixtures of conidia and two are for the green and yellow homokaryon controls. Wet a sterile wire loop with liquid medium and brush the loop lightly over the surface of the green parent culture to collect conidia; transfer these to the appropriately labelled bottle. Use three loopfuls of conidia to give a sufficiently high concentration. The loop is then re-sterilized and the same procedure carried out with the yellow-spored strain. For the mixture bottles add green and yellow conidia in about equal proportions, i.e. about two loopfuls of each.
3 Cap the bijou bottles and then agitate them on a whirlimixer, or shake them vigorously by hand. Leave the bottles overnight at 37°C. During this time the conidia will germinate and form a mat of mycelium on the surface of the medium. In the mixture bottles hyphal fusions will occur and many parts of the mycelium will contain nuclei of both genotypes.
4 On the following day, remove the mat of mycelium from one of the mixture bottles with a wire loop and spread out the mycelium on a plate of water agar. Let the surplus medium drain into the agar and then divide the mycelium into four pieces and transfer them to four places around the edge of a fresh plate of MM. Tease out the mycelium slightly to allow the heterokaryotic parts to grow

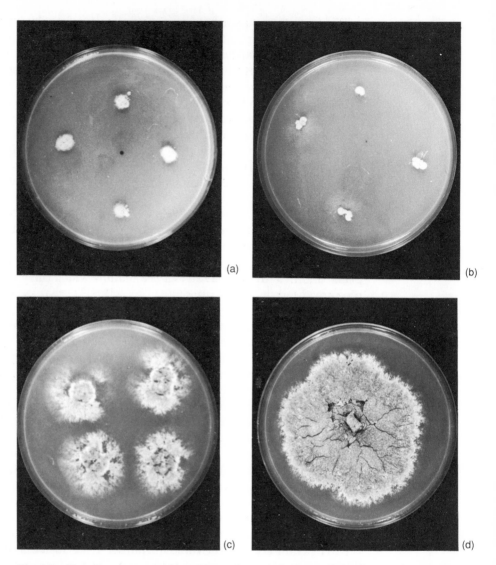

Fig. 6.3 Homokaryon controls and heterokaryons between the yellow and green strains of *A. nidulans* growing on minimal medium. The controls in (a) and (b) are the yellow and green auxotrophic strains inoculated as pieces of mycelium originating from an overnight growth of conidia on the surface of a complete liquid medium in bijou bottles at 37°C. In (c) the inoculum was mycelium from an overnight growth of a mixture of conidia from the yellow and green strains, and the vigorous growth of the heterokaryon can be clearly seen. The established heterokaryon in (d) is a sub-culture of mycelium from the starting culture in (c).

out freely. Repeat for the second mixture bottle, and then for the two control bottles, giving four culture plates in all.

5 Incubate the plates for 3–4 days and observe the vigorous growth of the hetero-karyons and the virtual absence of growth in the two control plates (Fig. 6.3). At this stage the cultures can be 'held' until the next practical class, or any

other convenient period within 2 or 3 weeks, by keeping them in a refrigerator at about 4°C.

6 Sub-culture the heterokaryons onto fresh plates of MM, making as many as required. To do this cut a $1/2$ cm^2 block of agar with a scalpel and then place this block upside down at the centre of the fresh MM plate. Avoid taking conidia and take a *thin* block of agar (i.e. not to the full depth of the plate). Incubate the new cultures upside down for about 1 week to give actively growing heterokaryons (Fig. 6.3d).

Observing and understanding the heterkaryons

Heterokaryons can be observed under a good quality stereomicroscope using a range of magnifications. The first thing to notice is that the culture has a mixture of green and yellow conida, which is the visual evidence that the two kinds of nuclei from the two parent strains are actually present in the mycelium. You will also notice that there are about equal numbers of green and yellow chains of conidia and in most cases all of the conidial chains on one conidiophore are of the same colour. In only a few cases are there mixed chains on the same conidiophore and these are quite difficult to spot. It seems that within the mycelium the two kinds of nuclei are not completely mixed, but rather they are 'bunched', and there is a tendency for all of one kind or the other to move into the vesicle during formation of the conidiophore. Since 'green' types arise from pre-existing 'green' types, and likewise for 'yellow', it is not surprising that they are segregated to some extent in the mycelium. Individual conidial chains are always uniform, reflecting their origin from a single 'mother' cell.

Note that the heterokaryon is a mixture of two kinds of haploid nuclei and that it is able to grow on a MM by *enzyme deficiency complementation*. In other words, each kind of nucleus is able to supply the other's deficiency: the 'green' genotype provides for the synthesis of phenylalanine and the 'yellow' genotype provides for the synthesis of adenine. Indeed, we may say that the heterokaryon is 'forced' and is only able to grow on MM as a heterokaryon.

An interesting follow-up exercise is to sub-culture mycelium of the heterokaryon onto CM and observe its mode of growth under these 'un-forced' conditions.

Apart from demonstrating enzyme deficiency complementation, this exercise on heterokaryons provides a useful basis for discussion and further reading about the role of heterokaryosis in nature, and in relation to plant pathogenic fungi. It can be pointed out that fungal colonies growing in natural populations will be uniformly green, but there is the possibility for mixtures of nuclei being present in mycelia which determine many different aspects of fungal growth and physiology. Heterokaryosis is a feature of many different species of fungi and could be of particular importance in those cases where the sexual cycle is lacking. There is the possibility, too, that the balance of nuclear types in a heterkaryon is not necessarily 1:1 but may vary and be adapted to conditions of the substrate on which the fungus is growing.

It is also useful to examine hyphae from the edge of the growing cultures (either from homokaryons or heterokaryons) under the microscope by scraping off small pieces of mycelium and mounting them in a drop of cotton blue dye. In this way it will be possible to observe the structure and development of the conidiophores.

2 Diploids

In heterokaryons it is often observed that sectors which are uniformly green or uniformly yellow arise. These are in marked contrast to the mosaic of green and yellow seen over the rest of the culture. These sectors are diploid and they arise by spontaneous chromosome doubling in nuclei of the mycelium. The most commonly observed diploids are green heterozygotes having the genotype +/y, ad/ +, +/phen. If these are sub-cultured they will be found to have larger conidia and to be capable of stable growth on MM. On CM they are unstable and revert back to haploids, and in the process they produce beautiful yellow sectors which arise when the chromosome carrying the wild type allele (+) for spore colour is lost.

3 Sexual Reproduction and Linkage Analysis

Cultures kept at 37°C for 10 days or more will readily form ascospores. In the heterokaryon formed between + ad + and y + phen some nuclear fusions that take place to give diploid ascus initial cells will be between pairs of genetically *unlike* nuclei ('crosses'), while others will be between pairs of *like* nuclei ('selfs'). Fusions that are 'crosses' give us an opportunity to study genetic linkage with Aspergillus. The 'crosses' that we require can be found by removing the ascospores from a number of individual cleistothecia and then growing up a sample of them to see whether they segregate to give a mixture of green and yellow colonies, or whether all colonies in the sample are uniformly of one colour or the other. Since the contents of one cleistothecium all originate from a single binucleate cell (see Introduction and Background) all of the ascospores in a hybrid fruiting body will originate from cells of the genotype +/y, ad/+, +/phen.

The haploid products of meiosis from nuclei heterozygous at three loci will comprise eight different genotypes, which arise by the processes of independent segregation and/or crossing over, depending on how the genes are linked. The genotypes we expect to find are given in Table 6.1 as reciprocal pairs.

Table 6.1 Genotypes expected following meiosis in Aspergillus heterozygous for y, ad and phen, given as reciprocal pairs.

+	ad	+	}	parental types
y	+	phen		
y	ad	+		
+	+	phen		
+	ad	phen	}	recombinant types
y	+	+		
+	+	+		
y	ad	phen		

The relative frequencies of the spore types of Table 6.1 will depend upon whether or not the genes are linked and, if so, how close together they are. If we take a sample of ascospores, determine their genotypes, and then estimate the relative frequencies of the eight types, we can make a chromosome map showing the linkage relationships of the three genes. The schedule below gives the step-by-step procedure for doing this. It basically involves plating out ascospores onto a growth medium on which *all* spores will grow, and then sub-culturing a sample of them onto a set of differential media to determine the genotype of each spore.

Isolating and cleaning cleistothecia

One mature heterokaryon culture, produced according to the schedule given in Exercise 1, will give sufficient material for several students. Its cleistothecia will be better developed towards the centre of the culture and may be obscured by an overgrowth of conidia and pinkish-white Hülle cells. Using a binocular microscope each student should remove three large cleistothecia with a needle and transfer them to a plate of water agar for cleaning.

Clean the cleistothecia by rolling them along the surface of the water agar, leaving behind all the conidia, the Hülle cells and mycelial fragments. When clean the cleistothecia are shiny black in colour, like miniature cannon balls. The cleistothecial wall is inert and the only living parts are the 300 000 or so ascospores contained inside.

Releasing the ascospores

To release the ascospores from their asci, first pick up the cleaned cleistothecia with a needle and transfer each one of them to the inside of a separately numbered bijou bottle containing 0.3 ml of sterile water. The water should have a drop of Tween 80 detergent (0.01% w/v) added to it before sterilization, as a wetting agent to reduce surface tension. It is important to place each cleistothecium on the inside of the bijou bottle *above* the water line. Then using the needle, drag up a droplet of water to the cleistothecium and burst the cleitothecium by pressing it against the side of the bottle. The droplet of water should become red coloured as the mass of ascospores is released, and the droplet can then be let down into the bulk of the water. Ignore the debris of the cleistothecial wall. Cap the bottles and agitate them on a whirlimixer to break up the asci and to release the individual ascospores into suspension.

Initial testing of ascospores

Test the contents of each bottle to determine whether the ascospores are of hybrid or selfed origin. This is done on a single plate of complete medium by streaking out a loopful of the suspension across the agar as three lines, one for each bottle (being careful to sterilize the loop between samplings from different bottles). Incubate the plates for 48 h at 37°C and then observe the colours of the streaks to see which are hybrid in origin (show a mix of yellow and green) and which are selfed in origin (all yellow or all green). Over a 48 h period the original suspensions will keep perfectly well at room temperature, and when the hybrid ones have been identified they can be used in the next stage of the procedure.

Adjusting the concentration of spores

The spore concentrations in the bottles deriving from hybrid cleistothecia need to be estimated in order to make appropriate dilutions for plating out the spores.

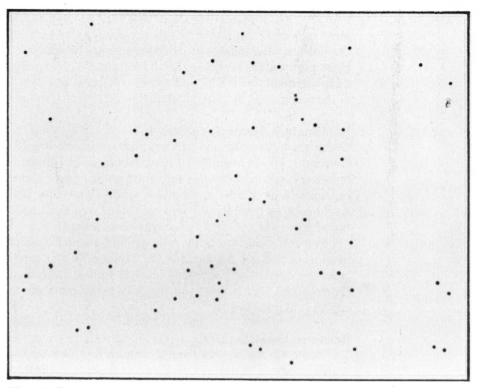

Fig. 6.4 Photograph of a suspension of ascospores of *A. nidulans* on a haemocytometer slide (x 130).

This is done using a haemocytometer counting chamber. The use of the haemocytometer is fully described in Chapter 4, and the same procedure can be followed for Aspergillus ascospores as is used for yeast cells. Fig. 6.4 shows the appearance of ascospores when viewed under the microscope.

Make serial dilutions to give suspensions containing 3×10^3 ascospores/ml. The dilutions need only be approximate, so that the concentrations of spores that are determined can be rounded up or rounded down to a convenient number before the dilutions are made. When 0.1 ml of the suspension is taken into another bijou together with 0.9 ml of the Tween 80, a tenfold dilution is made: i.e. from say 1 000 000 down to 100 000 spores/ml. A second tenfold dilution then goes from 100 000 to 10 000 and so on. Various other combinations of the suspension and the Tween can also be worked out where the figures are not quite so convenient, e.g. 0.1 ml suspension plus 0.5 ml Tween gives a sixfold dilution, and so on. It is helpful to use the whirlimixer between making dilutions, and to make sure that the spores are shaken well into suspension before removing the sample that is to be used. This part of the exercise should be looked upon as an important part of the practical because it gives training in a standard microbiological technique.

Plating out spores

Each student should plate out 0.1 ml of the final spore suspension onto each of five plates of fully supplemented MM + sodium desoxycholate (0.08%). In doing

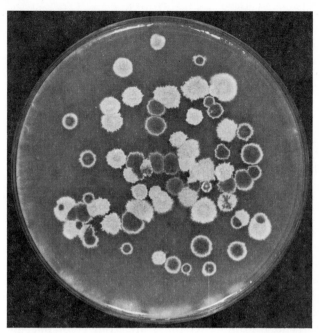

Fig. 6.5 Photograph of a fully supplemented MM plate showing colonies from a sample of ascospores. Note that the sample has about equal numbers of green and yellow colonies.

this, spread the inoculum with an L-shaped glass rod which is first sterilized by immersion in 70% alcohol followed by flaming (twice).

The above growth medium contains supplements of both adenine and phenylalanine so that all genotypes can grow. The sodium desoxycholate restricts the growth of the colonies and so keeps them separate from one another (Fig. 6.5). Since the germination level of the ascospores is only about 10% each plate should have in the region of 30 colonies after incubation for 3 days at 37°C, with each colony originating from a single ascospore. According to Mendel's First Law there should be segregation at the spore colour locus giving half green and half yellow haploid colonies and this can be confirmed by using the χ^2 test. At this stage the cultures can be refrigerated and kept until required; but handle the cultures carefully to avoid randomising the conidia among the different colony genotypes.

Determining colony genotypes

Each student should determine the genotypes of a random sample of 21 colonies from those growing on supplemented media, by replica-plating onto a set of differential media. In a class of less than 20 students it may be desirable to sample two or more sets of 21 colonies. The culture media required for the replica-plating procedure are as follows.

Plate of water agar
Plate of MM + adenine, i.e. the minus phen plate (-phen)
Plate of MM + phenylalanine, i.e. the minus ad plate (-ad)
Plate of CM

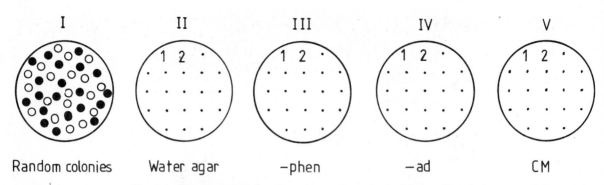

Fig. 6.6 Diagram showing the culture plates needed for replica plating onto differential media to determine the genotypes of a sample of 21 ascospores.

Also required is a template for marking the backs of the plates. The template can be the base of a petri dish with 21 holes burnt into it with a hot needle and spaced as shown in Fig. 6.6. The template is placed against the base of the media plate and the pattern transferred with a hot needle, and the needle marks are then spotted over with a felt pen so that the marked places can be seen through from the working surface of the plates. The four plates of media, together with the plate carrying the original colonies, are arranged on the bench in the order shown in Fig. 6.6. The plates in the series are given a reference mark to indicate where marked places 1, 2, etc. can be found throughout the series. Proceed as follows.

Sterilize a needle and then load it with conidia by rolling it about in one of the colonies of plate I. Clean the needle of excess conidia by stabbing it five times into marked place 1 on the water agar plate; and then use the needle to inoculate, by a single stab, marked place 1 of plates III, IV and V. When making the inoculation it is best to hold the plate upside down and stab upwards with the needle to allow surplus conidia to fall away from the agar surface.

The needle is then sterilized again by heating in a flame and a second colony is replica plated at marked place 2 on plates III, IV and V in the same way.

In order to take a *random sample* of 21 colonies it is important to take inocula from *every separate colony* on the dilution plate, and to sample from all five plates until the required number of colonies has been obtained. In a class of students this procedure will result in about equal numbers of yellow and green colonies being taken in an unbiased manner, which is important.

The water agar is just a working plate. Thus plates I and II are now discarded, while the other three are incubated at 37°C for 3 days to allow colonies to grow, or otherwise. Care is needed at this stage to obtain cultures that can be satisfactorily interpreted once they have grown (Fig. 6.7). The CM plate is a control to confirm whether a particular inoculation was successful or not, and also to confirm the colour phenotypes in cases where there is no growth on either of the other two plates. On all three plates III, IV and V the colours of the colony at any one marked place should be the same. If they are not, that colony should be excluded from the final classification. Some error is to be expected due to movement of conidia and due to the fact that some of the colonies in plate I (Fig. 6.5) may have originated from more than one ascospore.

The genotype of each colony sampled can be read by looking across plates III, IV and V (Figs 6.6, 6.7). If colony 1, for example, shows green on V, fails to grow

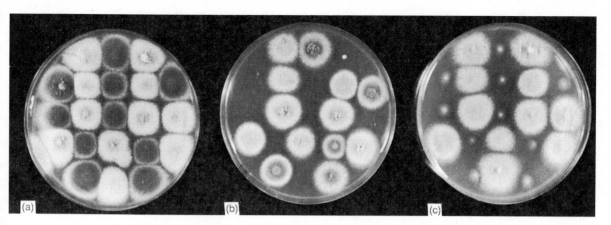

Fig. 6.7 Photographs showing the outcome of replica plating a sample of 21 colonies on a set of differential media to classify genotypes of ascospores. The plates are (a) complete medium, (b) -phen and (c) -ad.

on IV and gives good growth on III, then its genotype is **+ ad +** , and so on.

A typical set of class results for this gene mapping exercise in Aspergillus is shown in Table 6.2.

Table 6.2 A typical set of class results from determination of the genotypes of a sample of ascospores of *Aspergillus nidulans*.

Ascospore genotypes			Numbers
+	ad	+	140
y	+	phen	148
y	ad	+	7
+	+	phen	12
+	ad	phen	136
y	+	+	144
+	+	+	6
y	ad	phen	7
Total			600

Analysis and interpretation of results

Recall that the parent strains used were **+ ad +** and **y + phen** and the diploid heterozygous nuclei from which the ascospores originated were **+/y, ad/+, +/phen**. Note that the order of the genes given at this stage is abitrary.

To carry out the recombination analysis the genes are considered in all their possible combinations. Consider the **y–ad** pair first, by ignoring the **phen** column. The parental (original) combinations are **+ ad** and **y +**, and the recombinants are **+ +** and **y ad**. Secondly, for the **ad–phen** pair the parental combinations are **ad +** and **+ phen**, and the recombinants are **ad phen** and **+ +**. Finally, for the **y–phen** pair the parentals are **+ +** and **y phen**, and the recombinants **y +** and **+ phen**.

Using the numbers recorded in Table 6.2 we can now summarize the recombination data for all three pairs of genes, as shown in Table 6.3.

Table 6.3 Summary table showing numbers of recombinants for all pairwise combinations of genes, together with the totals for each pair and the recombination frequencies, derived from the data of Table 6.2.

	y–ad recombinants	ad–phen recombinants	y–phen recombinants
	12	136	12
	7	144	7
	6	6	136
	7	7	144
Total	32	293	299
% of total progeny (600)	5.3	48.8	49.8

In Table 6.3 the recombination values for pairs of genes have been calculated by expressing the recombinants as a percentage of the total progeny. These recombination values are then used to determine the linkage relationships of the three genes and to construct a linkage map, on the basis that 1% recombination is equal to 1 map unit. The low (5%) recombination frequency for the **y–ad** pair of genes indicates that these genes are closely linked (they are located in chromosome I). On the other hand, the near 50% recombination frequency for each of the **ad–phen** and **y–phen** pairs indicates that the **phen** locus is segregating independently of the other two and is thus located in a different linkage group (it is in chromosome III).

It is instructive to note from this exercise the difference between the procedure used for gene mapping in a haploid organism, such as Aspergillus, and that used in a diploid, such as Drosophila (Chapter 3). In haploids we determine the genotypes of the products of meiosis *directly*, whereas in diploids we work with phenotypes and have to resort to testcrosses to find the genotypes of the gametes giving rise to the testcross progeny.

REFERENCES

Fincham J.R.S., Day P.R. and Radford A. (1979). *Fungal Genetics*. California, University of California Press.

Clowes R.C. and Hayes W. (Eds.) (1968). *Experiments in Microbial Genetics*. Oxford, Blackwell Scientific Publications.

Chapter 7 POPULATION AND EVOLUTIONARY GENETICS

The exercises in this chapter are appropriate for all levels of study, from senior school to university undergraduate. The plant material which is used is both safe and easy to handle. The only precautions concern testing for cyanogenesis in white clover, where both picric acid (which is explosive when dry) and linamarin (which is poisonous) are used. These chemicals must be used with care. Picric acid must *always* be kept wet and care must be exercised both when transporting the liquid and when removing the top of the bottle (since liquid may have dried around the top). Sharp kitchen knives are used in the Brassica exercise so that care is needed here to avoid injury.

INTRODUCTION AND BACKGROUND

The understanding of evolutionary processes is an interesting and challenging area of biology. One way in which geneticists approach the issues involved is on the micro-evolutionary level, by investigating gene frequencies and the various forces which operate to determine or change these gene frequencies in populations. Change in gene frequencies is evolution. In this chapter we describe exercises with white clover which are both convenient and useful for the practical study of gene frequencies in populations.

Another approach to the practical aspect of evolution is to study 'human guided' evolution, and to observe the marked divergence that has occurred between the ancestral and present day forms of some of our domesticated plants and animals. Here we study divergence and evolution for quantitative characters amongst several forms originating from the wild cabbage species, *Brassica oleracea*.

Polymorphism and Cyanogenesis in White Clover

White clover (*Trifolium repens*) is polymorphic for two genes involved in its ability to produce small quantities of hydrogen cyanide, a process known as cyanogenesis.

Genetic polymorphism is said to occur when two or more distinct forms of a gene (or a species) are found in the same locality at the same time, and in such frequencies that the rarest form cannot be accounted for by recurrent mutation. The ABO blood group series in humans (Chapter 9) is a well known example of a genetic polymorphism. For a polymorphism to exist some force other than mutation must be operating to maintain the rarest allele concerned in higher than mutation frequencies. Natural selection is one such force (genetic drift is another), and by studying polymorphisms we are able to gain information about the way selection works to maintain or change certain gene frequencies in populations. Population geneticists use the term gene frequencies to mean allele frequencies (i.e. the relative frequencies of two or more alleles of a gene).

Cyanogenesis in white clover is controlled by two genes. One of these (symbolized **Ac**) controls the synthesis of two *cyanogenic glucosides,* from which HCN is produced by hydrolysis. When the dominant allele is present (genotypes **Ac/Ac** and **Ac/ac**) the glucosides are produced, but in the recessive homozygote (**ac/ac**) they are not produced. The other gene (symbolized **Li**) controls the production of an enzyme *linamarase,* which catalyses the hydrolysis of the glucosides. The enzyme is produced in the presence of the dominant allele (genotypes **Li/Li** and **Li/li**), but is not produced by the recessive homozygote **li/li**.

Combinations of these four alleles produce nine different genotypes. As summarized in Table 7.1, these produce four phenotypes when we consider dominance effects and the presence/absence of glucoside and linamarase enzyme. One of these phenotypes is strongly cyanogenic and another one is mildly so, while the other two are non-cyanogenic. Thus plants that possess both enzyme and glucoside (phenotype **Ac Li**) are strongly cyanogenic, because of enzymatic liberation of HCN from the glucoside. Plants that possess the glucoside but not the enzyme (phenotype **Ac li**) are mildly cyanogenic, because of slow, spontaneous (non-enzymatic) hydrolysis of the glucoside. Plants without the glucoside (phenotypes **ac Li** and **ac li**) are non-cyanogenic.

Table 7.1 Genotypes and phenotypes of cyanogenic clover plants. In the genotype **Ac/-, Li/-**, for example, the - means either the dominant or the recessive allele.

Genotype			Phenotype		
		Symbol	Glucoside	Linamarase	Cyanogenesis
Ac/Ac, Li/Li Ac/ac, Li/Li Ac/Ac, Li/li Ac/ac, Li/li	= Ac/-, Li/-	Ac Li	Present	Present	Strongly cyanogenic
Ac/Ac, li/li Ac/ac, li/li	= Ac/-, li/li	Ac li	Present	Absent	Mildly cyanogenic
ac/ac, Li/Li ac/ac, Li/li	= ac/ac, Li/-	ac Li	Absent	Present	Non-cyanogenic Absent
ac/ac, li/li		ac li		Absent	

The biochemical pathway of cyanogenesis is summarized in Fig. 7.1. In the **Ac Li** phenotype both glucoside and linamarase are present and thus enzymatic liberation of HCN takes place. In the **Ac li** phenotype this process is blocked at step 2, because of the absence of linamarase; in **ac Li** it is blocked at step 1 by the absence of the enzyme involved in the formation of the glucosides; and in **ac li** it is blocked at both steps.

The four phenotypes of Table 7.1 can be distinguished from one another by simple chemical tests, which involve additions of glucoside or linamarase

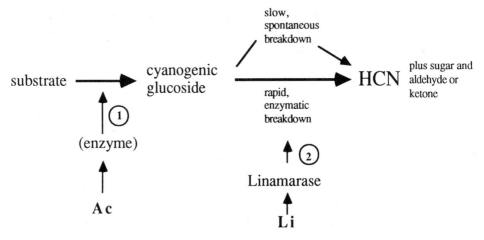

Fig. 7.1 Diagram showing the enzymatic formation of cyanogenic glucoside in white clover and its subsequent spontaneous or enzymatic breakdown to liberate HCN. The two gene loci involved are symbolized **Ac** and **Li**.

enzyme to crushed leaves and the use of sodium picrate paper to detect the liberation of HCN.

Under normal physiological conditions cyanogenesis is prevented from occurring in **Ac Li** plants, because the glucoside and linamarase enzyme are kept apart from each other. But under adverse environmental conditions such as frosting, which damages tissue and cells, glucoside and enzyme are brought together to cause cyanogenesis. Cyanogenesis damages the plant, probably because HCN inhibits respiration. Hence there is selection against the dominant **Ac** and **Li** alleles in populations in colder conditions such as those at high altitudes, which leads to relatively high frequencies of the recessive alleles in these populations. However, the dominant alleles do not disappear completely, because balancing selection operates to maintain them by discouraging grazing by small herbivores like slugs and snails, which dislike the HCN. In warmer climates where there is no frost, therefore, we find more of the dominant alleles. A study of cyanogenesis in clover, therefore, provides an example of how *balancing selection* operates in nature to maintain a genetic polymorphism.

Selection by Humans in Domesticated forms of *Brassica oleracea*

The wild form of *Brassica oleracea* is still found growing naturally in parts of Europe, where it is commonly referred to as the rock cabbage. At maturity it has a shortish stem with few leaves, an elongate terminal inflorescence and side shoots that also develop inflorescences (Figs 7.2a and 7.3a). The genetic variation within this single species has been manipulated by humans to produce several horticultural varieties which differ markedly in their heritable characteristics and which are all interfertile (Figs 7.2 and 7.3, b–g). In the cabbage, selection has been for a short stem and large scale production of overlapping leaves tightly surrounding the terminal bud. In Brussel sprouts, the emphasis has been for a greatly elongated stem and large scale production of lateral (axillary) buds (the sprouts). In kale, leaf production is accentuated but the habit is open and the

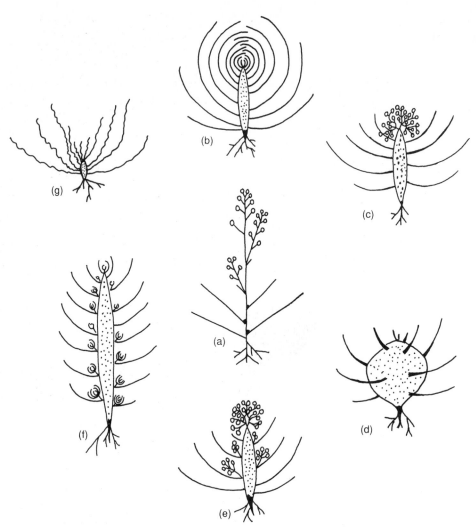

Fig. 7.2   Diagrams illustrating the main features of the original (undomesticated) and domesticated forms of *B. oleracea*. In the undomesticated form (a) no part of the plant is especially strongly developed. There are few leaves, the terminal bud has produced an elongate inflorescence with relatively few flowers, and two axillary (lateral) buds also have produced inflorescences. Three other axillary buds are shown. In the domesticated forms various parts of the plant have become strongly developed as a results of human selection: (b) the cabbage (terminal bud and associated leaves are greatly enlarged); (c) the cauliflower (terminal inflorescence enlarged); (d) kohl rabi (stem); (e) broccoli (terminal and lateral inflorescences); (f) Brussel sprouts (axillary buds); (g) kale (leaves; eaten or used decoratively). In each of b–g, non-edible portions are also enlarged to some extent, to allow for growth of those parts that are utilized as food. Stems are indicated by stippling, roots as solid lines, flowers as circles. Drawings not to scale.

leaves are curly. In the cauliflower we eat an enlarged, compact inflorescence consisting of thousands of fleshy flower buds. The broccoli is similar to cauliflower except that the inflorescence stalk is more elongate and fleshy, the inflorescence less compact and the flowers are usually green instead of white

Fig. 7.3 Photographs of wild and domesticated forms of *B. oleracea*, approximately $\frac{1}{10}$th natural size. (a) Wild form (European rock cabbage); (b) cabbage; (c) cauliflower; (d) kohl rabi; (e) broccoli; (f) Brussel sprouts; (g) kale.

(cream). Also, as well as the main terminal inflorescence numerous side inflorescences are produced from axillary buds in broccoli (which is therefore often referred to as 'sprouting' broccoli). In kohl rabi, which is perhaps the most remarkable form of the species, we harvest an enlarged stem packed with food reserve (and from the sides of which the leaves emerge in rather curious fashion).

The variation in the domesticated forms of *B. oleracea*, therefore, provides excellent material for the study of the effects imposed by humans on domesticated species. It also provides good evidence for the theory of evolution by natural selection.

Fig. 7.4 Some of the equipment needed for testing for cyanogenic phenotypes of clover. Distilled water, toluene and linamarin are shown in small dropper bottles; the large dark bottle contains picrate paper strips. The plastic bag on the left contains a clover piece sampled from a population. Vials, cotton buds, a glass rod for macerating leaf tissue, tie-on labels and a potted **ac Li** (enzyme) plant are also shown.

PROCEDURES

For Studying Cyanogenesis in White Clover

Equipment required

For sampling a population and distinguishing cyanogenic phenotypes

Medium sized plastic bag (1 per student),
Small plastic bags (1 per plant sampled from the field),
Tie-on labels,
Small glass vials (say 3 x 1 cm) with caps or (ideally) rubber stoppers,
Flat bottomed glass rods that will fit inside the glass vials,
Cotton buds,
Dropper bottle with distilled water,
Dropper bottle with toluene,
Dropper bottle with 0.1% linamarin (or potted **Ac li** plant; see below),
Picrate paper strips in dark bottle (see below for preparation),
Incubator set at 37°C (optional),
Rubber bands or wooden blocks to hold vials,
Potted **ac Li** plant.

Some of the equipment listed above is shown in Fig. 7.4.

Fig. 7.5 Aquarium containing clover plants, two each of **Ac Li, Ac li, ac Li** and **ac li** phenotypes, about to be browsed by snails (on inside of lid). The pots have been well surrounded with soil and thoroughly watered.

For setting up a browsing experiment
Medium-sized aquarium with glass lid (Fig. 7.5),
One or two potted plants of cyanogenic and non-cyanogenic clover,
Soil or equivalent material,
Snails.

Preparation of picrate papers and linamarin
Picrate papers are used for determining the cyanogenic phenotypes of clover, and are made by soaking filter paper in a solution of sodium picrate and then hanging the paper up to dry. When dry, the paper should be cut into manageable strips about $\frac{1}{2}$ cm wide, and the strips should be stored in a dry, dark bottle to prevent discoloration by light and moisture. It is best to make fresh strips each year.

Sodium picrate is made by neutralizing a saturated solution of picric acid with sodium bicarbonate; or by dissolving 5 g solid (*but moist*) picric acid* and 25 g sodium carbonate in 1000 ml distilled water. Filter the prepared solution.

Linamarin can be purchased from the usual biochemical supply houses as a powder (check the price before ordering!). Mix it in *small* quantities in a dropper bottle with distilled water to make the 0.1% solution. Store at 4°C or thereabouts (in a refrigerator).

*Note that picric acid is explosive when dry and hence must *always* be kept wet. Exercise care both when transporting the liquid and when removing the top of the bottle (since liquid may have dried around the top).

Fig. 7.6 Equipment used for studying the effects of human selection in *B. oleracea* . Roots, leaves, inflorescences and stem (on the pan of the scales) of a single specimen of broccoli have been separated and are about to be weighed.

Obtaining plants of known phenotype

These can be obtained as cuttings by sampling a population using the tests described below. Use the 'slow release' test [step 3] to locate **Ac li** plants; and then use these to locate **ac Li** plants [step 2]. Once obtained, cuttings of known phenotype can be grown easily in pots: though be wary of aphids and slugs, and keep different phenotypes well separated so that they do not become mixed.

For Studying Selection by Humans in *Brassica oleracea*

Equipment required

The equipment required here is very modest. The most important item is a simple, top-pan, kitchen style balance weighing up to a few pounds (kg) (Fig. 7.6); one per student or group of three or four students is desirable. In addition, a supply of reasonably sharp kitchen knives is necessary, as well as a large container for the disposal of unwanted plant material (most edible parts are either consumed immediately or gladly taken home for an evening meal!).

Obtaining material

Apart from the wild form, the varieties of *Brassica oleracea* required are cabbage, cauliflower, broccoli (sprouting broccoli), sprouts (Brussel sprouts), kohl rabi and kale (kitchen or curly kale). Seedlings of these varieties can be grown easily in trays or pots. In the case of the original form of *B. oleracea*, seed can be collected from wild populations (with very special consideration to conservation issues) or obtained from an appropriate supply house.

Mature (harvest stage) specimens can be grown from purchased seed or seedlings. With this approach it is necessary to check maturation times beforehand,

and then to plant seed or seedlings so as to obtain specimens all of which have matured together. Plants do best if grown outside rather than in pots in a glass-house. Alternatively, plants can be obtained from a market gardener, by arranging for entire plants, *including roots*, to be dug up and rid of soil. A combination of the two approaches may be necessary. Uprooted plants can be held for a few days in plastic bags in a refrigerator or cooler. Note that retail shops are a poor source of material for this exercise since their plants normally have been stripped of un-utilized parts.

EXERCISES

1 Frequencies of Cyanogenic Phenotypes in Natural Populations of White Clover

This exercise involves sampling a natural population(s) of white clover followed by determining in the laboratory the cyanogenic phenotype of each plant collected. Data are then analysed, compared with others and then interpreted in population genetic terms.

Choosing and sampling from a suitable population

A suitable population should be found beforehand and should be well established and undisturbed. The borders of forest and grassland or areas with scattered scrub growing over grass and herbs usually have good populations of clover. It is useful to obtain from meteorological sources the altitude and mean winter temperature and/or mean number of frosts for the population in question. If more than one population is sampled these can be chosen to cover a range of environments with different altitudes and frost level, for example.

Sampling can be done either by collecting seed from the population in question and producing seedlings for testing; or collecting stolons (creeping stems) of adult plants and testing these. The latter approach is less time consuming and can be done without concern as to the flowering of clover, which is sometimes rather erratic.

Working from seed

If working from seed, collect a random sample of mature seedheads. Allow these to dry completely in paper bags before rubbing the heads vigorously to free the seed from their pods.

Clover seed are rather slow to germinate. It is best to scarify (weaken) the seed coat first by immersing the dry seed in concentrated sulphuric acid for 15 min, followed by washing with plenty of water. Germinate the seed on moist filter paper in large petri dishes or plastic containers with lids.

Sampling a population of adult plants

This will take between 1 and 3 hours depending on how extensively the population is sampled and how far it is from the laboratory. Usually students (and teachers) enjoy this part of the exercise! When in the field it is desirable to demonstrate how to identify white clover from similar plants (red clover, *Oxalis*, *Medicago*, etc.); and how to collect a plant piece.

To sample the population, students should be well spread out and plants sampled 3 m apart (to avoid sampling more than once from the same plant) in rows. Collect stolons each with at least three or four mature, healthy leaves, after first carefully clearing each plant of surrounding vegetation so as not to remove or damage the leaves (which is especially important if the plants are also to be scored for browsing; see below). Collect stolons from an appropriate number of plants (10 per student is convenient for a class of 25), placing them in a plastic bag for transport back to the laboratory.

On returning to the laboratory, number each plant with a tie-on tag and place each plant in its own separate plastic bag. In the bottom of each bag place a small amount of water before filling the bag with air and tying it shut. Plants will then remain healthy for 4 or 5 days if kept in a cool place.

Distinguishing cyanogenic phenotypes
Up to ten plants can be analysed together without difficulty, many more if only step 1 is carried out. Hold the vials for each step of the analysis together with a rubber band or place them in a rack made by drilling holes of the appropriate size in a block of wood (Fig. 7.7). Keep careful records in a laboratory book as each step is set up and completed.

Step 1: Identifying Ac Li plants from all others
Take one fresh leaf (that is, three leaflets), tear it into small pieces and place the pieces in a small glass vial. Add 1 drop of distilled water and macerate the leaf material with a flat bottomed glass rod. Add 2 drops of toluene and mix the contents. Clean off any tissue or liquid from the sides of the vial with a cotton bud. Re-bag the specimen.

Label a 5 cm strip of picrate paper with the number of the plant being tested and place the strip into the vial, folding it over the lip such that the paper does not come into contact with the contents of the vial. Cap the vial firmly (a rubber bung is best).

Incubate the preparation for 2 hours at 37°C (or 24 hours at 20°C). Examine for colour change (yellow to red-brown) and record results. A colour change indicates that the leaf contained both glucoside and enzyme and was thus of phenotype **Ac Li**.

The analysis may be halted at this point if desired, by regarding all those plants that give no colour change with the above test as non-cyanogenic. This loses a certain amount of useful information, but is appropriate when the following techniques are difficult or too time consuming.

For the identification of all four cyanogenic phenotypes, plants giving no colour change after step 1 are tested further, as follows.

Step 2: Identifying ac Li plants
This is done by adding linamarin to plants that do not give positive reactions after step 1. As soon as possible after completion of step 1, remove the picrate paper from any vial that has not been identified as **Ac Li**. Add 2 drops of 0.1% linamarin. Mix the contents, clean the sides of the vial again if necessary and supply a fresh, labelled, picrate paper strip. Recap the vial and incubate as before. Record any colour change. If the enzyme linamarase is present it will react with

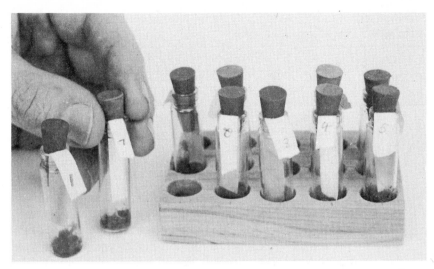

Fig. 7.7 Results of testing leaves of ten clover plants for **Ac Li** phenotype (2 hours incubation at 37°C). Vial 1 (left) shows a positive test for HCN production; vial 7 is negative.

the glucoside to give a colour change to the picrate paper.

If step 2 does not follow soon after step 1 has been completed, use *fresh* leaf material from the unknown plant being tested. This is because enzyme activity decreases sharply 3–4 hours after first macerating a leaf.

If pure linamarin is unavailable, use one fresh leaf of a plant known to be **Ac li** in phenotype as a linamarin source, macerating and mixing it well with fresh leaf material from the unknown plant.

Step 3: Identifying Ac li plants

Ac li leaves give a *slow*, mild colour change to a picrate paper after 24 hours incubation at 37°C, because of spontaneous liberation of HCN (see Fig. 7.1). However, this test is sometimes unreliable for identifying all **Ac li** plants. A better test involves adding linamarase enzyme to plants that give no colour change after step 2. Leaf material from a plant known to be **ac Li** is the best source of the enzyme. The test is done by combining a fresh leaf from an unknown plant from step 2 with one from an **ac Li** plant, macerating and mixing the leaf material together. If the glucoside is present in the leaf from the unknown plant it will be reacted on by the added enzyme to give a colour change to the picrate paper.

Plants that give no colour change after steps 2 and 3 must lack both glucoside and linamarase and must therefore be of phenotype **ac li**.

Steps 2 and 3 may be done in reverse order if desired.

Analysis of data and comparisons

Pool results over the whole class to obtain a good sized sample. Express the frequencies of the four (or two) phenotypes as percentages of the total plants tested. Actual data are shown in Table 7.2.

Comparisons can be made with data from different populations experiencing different frost levels or growing at different altitudes, for example. These can be

Table 7.2 Percentages of four clover phenotypes collected as plant pieces from three Wellington (NZ) populations experiencing different average frost levels.

Population number	Mean ground frost (days in July)	Altitude (m)	Ac Li	Ac li	ac Li	ac li	No. of plants
1	7.8	15	74	17	6	3	178
2	10.2	65	50	23	15	12	155
3	14.0	56	36	33	12	19	176

from previous studies of the class or from published work. Statistical comparisons can be made using contingency chi-squared tests (see Appendix); and visual comparisons can be made by presenting the data as histograms or pie charts. These will illustrate the relationship between phenotype and frost levels, and phenotype and altitude (Tables 7.2 and 7.3; Fig. 7.8).

Table 7.3 Percentages of four phenotypes in natural populations of clover from different altitudes of the Central European Alps; adapted from Daday (1954).

Altitude (feet)	Phenotype				No. of plants
	Ac Li	Ac li	ac Li	ac li	
1903	71	15	11	3	184
2296	28	53	2	17	100
3510	11	40	6	43	110
4593	1	12	4	83	90
5577	0	9	7	84	94
6398	0	0	4	96	99

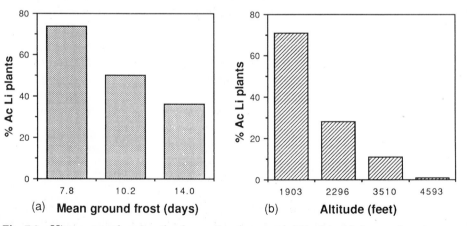

Fig. 7.8 Histograms showing the decreasing frequency (%) of **Ac Li** clover phenotypes with increasing frost levels (a) and increasing altitude (b). Data from Tables 7.2 and 7.3, respectively.

2 Selective Browsing of Cyanogenic Phenotypes of White Clover

Selective browsing of white clover by small herbivors can be demonstrated in the laboratory by offering snails cyanogenic and non-cyanogenic plants as food.

Setting up the experiment

Prepare an equal number of well established, potted clover plants of phenotypes Ac Li and ac li (or all four phenotypes for a more thorough experiment) by removing dead, dying and browsed leaves. Then trim the plants so that roughly equal numbers of leaves of the different phenotypes are present (50–100 per plant is suitable). Make sure that when removing a leaf it is cut off at the *base* of its stalk, so that it will not be confused later with a leaf that has been fully browsed. It is a good idea also to cut off or mark with nail varnish or paint all the terminal buds of each plant, so that newly produced leaves can be excluded from the analysis.

Place the plants in an aquarium such that no phenotype will be favoured because of placement. Pack soil or equivalent material all around and to the tops of the pots; and then place on the inside of the glass lid of the aquarium a number of snails (about two per pot) that have been starved for 48 hours. Water the plants well to encourage feeding, apply the glass top and leave the snails to browse at room temperature for 5–7 days (Fig. 7.5). Monitor the progress of the experiment daily. Halt the experiment before any one phenotype has been completely eaten.

Analysis of results

At the end of browsing, remove the plants carefully from the aquarium and examine them. If browsing has been intense the results may not require detailed analysis. If not, obtain a measure of the degree of browsing of each phenotype by giving each plant a 'browsing coefficient'. This is obtained by totalling the number of leaflets that have been partially or completely eaten and dividing this number by the total number of leaflets examined. A more thorough analysis is also possible by assigning each leaflet a score of 0 – 4 according to the following scheme.

Score	Degree of browsing of leaflet
0	not eaten at all
1	$>0 <1/3$ eaten
2	$>1/3 < 2/3$ eaten
3	$> 2/3<$ all eaten
4	eaten completely

Then the browsing coefficient is obtained for each plant by multiplying each score by the number of leaflets found with that particular degree of browsing, summing all such values obtained and then dividing by $4N$, where N is the total number of leaflets examined. For those more mathematically inclined this is:

$$\frac{\Sigma n_i b_i}{4N},$$

where Σ = sum, b_i = the ith browsing score (i from 0 to 4), n_i = the number of leaflets in the ith browsing category and N = the total number of leaflets.

Table 7.4 Browsing of leaves of four cyanogenic phenotypes of clover (two potted plants of each) by eight snails for 5 days.

Phenotype	Total no. leaves (leaflets)	Total no. leaflets browsed	Browsing coefficient
Ac Li	104 (312)	52	0.17
Ac li	111 (333)	80	0.24
ac Li	98 (294)	120	0.41
ac li	120 (360)	141	0.39

Typical data are shown in Table 7.4, from which it is evident that browsing was preferentially on **ac Li** and **ac li** plants. Fig. 7.9 illustrates the result of browsing by 12 snails for 7 days, which gives a clear visual demonstration of differential browsing.

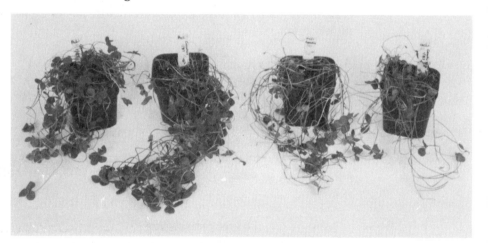

Fig. 7.9 Results of browsing of cyanogenic clover plants from the experiment set up in Fig. 7.5. The order of pots from the left is **Ac Li, Ac li, ac Li** and **ac li**. Browsing has been much stronger on the non-cyanogenic phenotypes **ac Li** and **ac li**.

Browsing in natural populations
Browsing can also be assessed on plants collected from natural populations. For this exercise, examine each plant for its browsing coefficient *before* it is tested for its cyanogenic phenotype, keeping separate records for each numbered plant. Obtain class result and derive an overall browsing coefficient for each phenotype separately by averaging. Compare browsing coefficients of cyanogenic versus non-cyanogenic plants.

The data of Table 7.5 were obtained by a class of 17 students (ten plants per student) from a natural population growing in an exposed (wind swept), habitat, at 280 m altitude. Each plant was first scored for browsing and then analysed for its cyanogenic phenotype. Note that in this natural population browsing is a lot lighter than in an experimental situation (cf. Table 7.4); but is also preferentially on non-cyanogenic phenotypes **ac Li** and **ac li**. Other natural populations may not show this differential browsing, perhaps because low temperatures are not conducive to selective browsing; or perhaps because the browsers are not snails and could not differentiate between the clover phenotypes.

Table 7.5 Numbers of plants and browsing coefficients of four phenotypes from a natural population of clover from Wellington (NZ).

Phenotype	No. of plants (%)	Total no. leaflets	Total no. leaflets browsed	Browsing coefficient
Ac Li	73 (43.0)	678	59	0.087
Ac li	42 (24.7)	474	42	0.088
ac Li	23 (13.5)	282	38	0.134
ac li	32 (18.8)	360	49	0.135

Contingency χ^2 (total browsed vs total un-browsed leaf material in cyanogenic vs non-cyanogenic phenotypes) =10.4, $d.f. = 1$, $P. < 0.01$, highly significant.

3 Interpretation of Cyanogenesis Results

After completing some or all of the above exercises an attempt should be made to explain why different populations of clover have different frequencies of cyanogenic versus non-cyanogenic phenotypes. Special attention should be given to the data that show a relationship between phenotype frequencies and altitude (temperature) and frost levels, such as those shown in Tables 7.2 and 7.3; and to data showing differential browsing of cyanogenic versus non-cyanogenic plants (Tables 7.4 and 7.5).

Some issues to consider are as follows.

1 In the healthy plant the cyanogenic glucoside and the enzyme linamarase are kept apart from each other, so that HCN is *not* liberated into living plant tissue.
2 Environmental stresses such as frosting and drought cause leaf damage, which brings together the glucoside and the linamarase enzyme.
3 HCN is an enzyme inhibitor and thus affects probably a wide range of metabolic processes in plant as well as animal tissue.
4 Temperature influences the frequency and distribution of herbivors; and may influence the degree of selective browsing by small herbivors.

These issues can be considered in the light of the current view that two opposing selective forces account for the polymorphism for cyanogenic phenotypes in clover populations. These are:

1 *Cold temperature and other high altitude stresses*, which induce HCN production in intact plants. These conditions favour, therefore, the growth (and presumably also reproductive fitness) of non-cyanogenic forms. Clover reproduces mainly by vegetative growth so that physiological fitness could be more important than reproductive fitness.
2 *Differential browsing of non-cyanogenic phenotypes by small herbivors*, which favours the growth of cyanogenic plants, especially in warmer regions where selective browsing may be strongest and where herbivors are likely to be more numerous and more active.

4 Human selection in *Brassica oleracea*

In this exercise domesticated forms of *B. oleracea* are dissected into their main parts, which are then weighed so as to come to an appreciation of the nature and extent of modification brought about by selection. The original form of the species is also examined, as are 3 – 6 week old seedlings of all the forms. The latter are used to review the main morphological features of the plant; and to indicate how very similar the various forms of the species are at a young age.

Handling material
Each student or group of students should first study a mature specimen of the original form of the species. If this is not possible the original form of another species of the genus such as *B. campestris* will suffice. Examine the material as follows.

1 Identify *roots, stems, leaves* and *axillary (lateral) buds* (some of which may have produced lateral stems); *inflorescence* with flowers, fruits and seed (or *terminal bud* if the plant has yet to produce flowers).
2 Carefully remove any remaining soil from the roots. Cut up the specimen into four parts, namely leaves, stems, inflorescence including its stalks and roots. Pool like parts into manageable lots, weigh each lot and record results.
3 Examine each domesticated form in turn, identifying the various morphological parts as before. Remove soil, cut up, pool and weigh as before, being especially careful to identify the true junction between root and stem (below the lowermost leaf base scar). *For cabbage, kale, Brussel sprouts* and *kohl rabi* there will, of course, be no inflorescence (unless the plants have bolted!). For *broccoli*, include small lateral inflorescences with the main flowering head, but add inflorescence leaves with the main leaves. For *cauliflower* and *broccoli*, include the inflorescence stalks with the flowers. For *cabbage*, cut the specimen in half longitudinally and note the terminal bud and tightly packed leaves. Cut off the root; and then the leaves by removing the stem as one piece. For *Brussel sprouts*, weigh sprouts and other leaves separately. Record results. If some forms are not available in good supply, demonstration plants or photographs provide a reasonable substitute, especially if students are supplied with representative data collected beforehand to work with.
4 Examine seedlings of various domesticated forms, noting their strong similarity.

Table 7.6 Weights of plant parts in grams and percentages of total weight (in brackets) in the undomesticated and domesticated forms of *B. oleracea*. N = number of specimens examined.

Form	N	Stem	Leaf	Root	Ax. bud	Inflor.	Total
Undom. form	2	40 (18)	78 (36)	82 (37)	-	20 (9)	220
Cabbage	4	430 (6)	6480 (89)	345 (5)	-	-	7255
Cauliflower	4	580 (6)	5290 (58)	621 (7)	-	2619 (29)	9110
Broccoli	4	776 (17)	2464 (53)	340 (7)	-	1040 (23)	4620
Brussel sprouts	4	1480 (19)	2400 (31)	660 (8)	3340 (42)	-	7880
Kohl rabi	3	471 (76)	117 (19)	33 (5)	-	-	621
Kale	3	91 (24)	226 (60)	60 (16)	-	-	377

Analysis and interpretation of results

1 Pool data for all plants of the one type. Obtain the total weight for each plant type and calculate the percentages contributed to this total by the four parts (Table 7.6). Treat Brussel sprouts and other leaves separately, appreciating the fact that the contribution of lateral buds to the total weight of all crops other than Brussel sprouts is essentially zero.

2 Construct pie charts from the percentage values, labelling the sectors appropriately (Fig. 7.10). Indicate for each type the part of the plant that has been developed by selection and directly utilised as food.

3 Compare the total weights and pie charts of the various forms of the species. Rank the domesticated forms according to their efficiency in relation to human use.

As a useful conclusion to this exercise attention may be directed to the following questions:

1 Are there any features present in the original form that are not expressed in any of the domesticated forms, and vice versa? Consider leaf hairiness and colour, for example.

2 Which is the most efficient and which the least efficient form, as far as immediate use by humans is concerned? Fig.7.10 is especially revealing in this respect! What other factors need to be considered when assessing overall efficiency?

3 If allowed to flower, all the various forms of the species can be hybridized to produce fertile F_1 and F_2 progeny. What conclusion is justified?

4 Since the full life cycle (to fruit and seed) is normally curtailed by harvesting, how are the domesticated forms perpetuated; and perpetuated as pure lines?

5 Usually more than just the utilized part of a crop becomes developed by selection (Table 7.6, Fig. 7.10). Why should this be so?

6 No form of *B. oleracea* in which the *root* is utilized as food has yet been successfully developed. Why might this be so? Note that the original form of the related species *B. campestris* has been so developed, to produce the turnip.

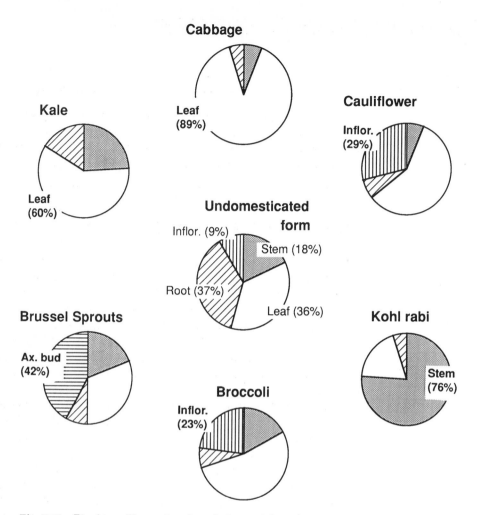

Fig. 7.10 Pie charts illustrating the relative weights of roots, stems, leaves, axillary buds and inflorescences of undomesticated and domesticated forms of *B. oleracea* . For the un-domesticated form (centre) the relative weights of the various plant parts are given as percentages of the total weight, thus also serving as a legend for the other forms. For each domesticated form, the part of the plant enlarged and utilized directly as food by humans is indicated in bold underlined type, along with the percentage of the total weight it represents. For all forms except Brussel sprouts the contribution of axillary buds to the total weight is essentially zero. Data from Table 7.6.

Researching the origins of each domesticated form of *B. oleracea*, as far as they are known, would make an interesting project.

REFERENCES

Daday H. (1954). Gene frequencies in wild populations of *Trifolium repens*. II. Distribution by altitude. *Heredity*, **8**, 377–384.

Richardson W.N. and Stubbs T. (1978). *Plants, Agriculture, and Human Society*. New York and Amsterdam, W.A. Benjamin.

Chapter 8   PLANT GENETICS

The exercises presented in this chapter are suitable for all levels of study, from senior school to university undergraduate; though teachers should select exercises most appropriate to their particular class situation. There are no problems with matters of safety or plant conservation in this chapter.

INTRODUCTION AND BACKGROUND

Studying genetics with plants is not as easy as it is with Drosophila, largely because of the long generation time of most plant species and the difficulties associated with maintaining genetic stocks and setting up crosses. However, seedlings of certain plants such as maize, sorghum, tomato etc. and maize ears provide excellent material for teaching genetics in a laboratory. Their only drawback is that normally only the F_2 generation is observed. However, this situation can be turned to advantage as part of the practical work if students are set the task of working out for themselves what the parents and F_1 were for each particular exercise.

Seedlings and maize ears can be used to illustrate a wide range of genetic phenomena, from simple Mendelian ratios to more complex ratios brought about by interactions between genes (epistasis), the full interpretation of which also introduces students to some principles of biochemical genetics. This material helps to provide balance to a genetics laboratory, which can easily become overburdened with Drosophila. Moreover, use of seedlings and maize ears makes it possible to demonstrate easily that the principles of genetics apply to a range of different organisms, such as maize, sorghum, peas, and tomato, and thus presumably to *all* organisms.

Genetics with Seedlings

A number of biological supply houses offer a good range of quality seed at reasonable cost for use in genetics classes. The seeds are germinated by students or teachers, and the seedlings so produced are scored for their various phenotypes and interpreted in terms of their origin and the laws of inheritance. The seedlings represent the F_2 derived from crossing two parent lines carrying the traits in question. The F_1 are self-pollinated and the seed produced made available for purchase by the supply company. The seedlings are examined when 2–4 weeks old. The characters in question have been selected precisely because they become apparent at the seedling stage of growth, and also because of their clarity and because they are not influenced greatly by growth conditions. The characters include the tall/dwarf traits of maize; the albino trait of sorghum and other species; and the hairy versus hairless state of the tomato.

Some of the main seedling types that can be grown from currently available seed are given below, grouped according to the pattern of inheritance they illustrate.

Monohybrid ratio with complete dominance

1 *Red* vs *non-red* (green) coleoptile (the protective sheath) and stem of sorghum. These two characters are easily distinguished in even very young seedlings, such that planting of seed to completion of the exercise need take no more than 2 weeks. Germination of 'seed' (strictly these are fruits!) is very good and even. Red is dominant to non-red.

2 *Tall* vs *dwarf* plants in maize and peas (Fig. 8.1a). The difference between tall and dwarf can be seen easily 1–2 weeks after the appearance of seedlings, especially in maize, simply by reference to seedling height. The sometimes erratic and uneven germination of maize can create problems in scoring young seedlings accurately, though the first leaf of dwarf plants, being more oval than in tall ones, is a very useful aid to identifying the dwarf phenotype. Tall is dominant to dwarf.

3 *Hairy* vs *hairless* leaves and stems in tomato (Fig. 8.1b). This is a very clear character difference, best seen in seedlings 3 or more weeks old by illuminating the seedling well against a dark background. Use of a small hand lens is helpful though not essential. Hairy is dominant to hairless.

4 *Cut* vs *potato leaf* in tomato (Fig. 8.1c). The difference is distinct from the first true leaf of the seedling on. Cut is dominant to potato leaf.

5 *Green* vs *albino* seedlings, especially in sorghum, maize and lettuce (Fig. 8.1d and g). The albino state is caused by a recessive allele which inhibits the formation of chlorophyll. This is a *lethal* condition in the post-seedling stage (after food reserve in the seed has been exhausted), making this material especially interesting. The albino seedlings are very distinctive, being pure white or sometimes pale yellow (depending on the amount of carotenoid pigment produced). They wither and die soon after germination. Green is dominant to albino.

Other useful, though less distinctive, maize characters segregating to give monohybrid ratios in seedlings are *green* vs *white sheath* (chlorophyll present or absent in the base of the leaf surrounding the stem); *ligule* vs *ligule-less* (the ligule being a small papery appendage found on the joint between leaf and leaf base, facing the stem); *normal* vs *glossy leaves* (in the latter a fine mist of water sprayed onto the leaf produces adhering droplets of water, rather than a uniform wetness).

In each of the cases listed above the character difference is determined by a pair of alleles, one of which is completely dominant over the other. Therefore the F_1 of a cross between the two forms are of one type (the dominant type); and the F_2 show a typical Mendelian 3 : 1 ratio. Because of its lethal nature, the albino trait is inherited only through heterozygotes which, when self-pollinated, produce a 3 : 1 ratio in their progeny.

Monohybrid ratio with incomplete dominance

6 *Green, yellow-green, yellow leaves* in soybean, tomato and tobacco (Fig. 8.1e). The yellow seedlings are distinguishable soon after germination and should be

Fig. 8.1 2–4-week-old seedlings of various types used in teaching practical genetics. (a) Maize, tall and dwarf (arrow); (b) hairy (right) and hairless tomato; (c) cut (i.e. dissected) (right) and potato leaf tomato; (d) plastic tray containing sorghum seedlings segregating 3 : 1 for green and albino; (e) green (left) and yellow-green tomato. Yellow seedlings have already died; (f) maize. From right to left: tall, green; tall, albino; dwarf, green; dwarf, albino; (g) sorghum, segregating 9 green : 7 albino (9 : 3 : 4 if the two albino conditions are distinguished).

scored immediately, for they will quickly die. Green and yellow-green seedlings are distinguishable a few days later. As with the albino trait in sorghum and maize, the yellow-green and yellow conditions are caused by a chlorophyll deficiency. In these cases, however, the heterozygotes show an intermediate phenotype (yellow-green) because the relevant alleles are *incompletely* dominant. Yellow-green seedlings survive to flowering and are self-pollinated to perpetuate the segregating line. The F_2 therefore shows a 1 : 2 : 1 ratio of the three types.

Dihybrid ratio with complete dominance

7 *Tall* vs *dwarf* and *green* vs *albino* in maize (Fig. 8.1f). This is a combination of the character pairs described in (2) and (5) above. Thus four phenotypes are found in F_2 seedlings. The two pairs of alleles concerned are in separate chromosomes, which therefore segregate independently during meiosis of the F_1 to produce a standard 9 : 3 : 3 : 1 ratio in the F_2. The somewhat erratic germination of maize can create problems with obtaining sufficient data to analyse, but simply showing students the progeny types and providing them with data obtained previously still represents a worthwhile exercise.

8 *Green* vs *albino* and *red* vs *non-red* in sorghum. This is a combination of the character pairs described in (5) and (1) above. As for the material in (7), four phenotypes appear in the F_2 in a 9 : 3 : 3 : 1 ratio. The red trait is easily seen in both green and albino seedlings.

Dihybrid ratio with interacting genes (epistasis)

9 *Green* vs *albino* in sorghum and maize. In both of these species, *two* distinct albino genes are known, located in different chromosomes. Separately each gives a 3 : 1 ratio in the F_2, as in (5); but when combined in the one cross three of the expected four phenotypic classes are albinos. Thus an F_2 ratio of 9 green : 7 albino results. Trays of seedlings showing, on the one hand, a green : albino ratio of 3 : 1 and, on the other hand, a 9 : 7 ratio make for very interesting study.

Genetics with Maize Ears

Maize (*Zea mays*), also known as corn, has several features which make it particularly suitable for the study of genetics. Because of this, and because of its agricultural importance, the genetics of maize has received probably more attention than any other plant species.

Maize is a monocotyledonous plant and a member of the grass family (Gramineae). Cultivated forms grow to about 2 m in height and usually have a single stem with leaves arising from their internodes (Fig. 8.2a). Male flowers and female flowers develop from separate parts of the plant. The ear (female) develops as a branch in the axil of one of the leaves about midway up the stalk, and is composed of several hundreds of female flowers, each with its own long silk (the style with its stigma). The silks hang out from the top of the ear and can receive pollen along the whole of their lengths. The male flowers (tassels) are carried at the top of the plant. When the pollen is ripe it can be used in experiments for crossing onto the silks of a different plant, or selfing by application to the silks of

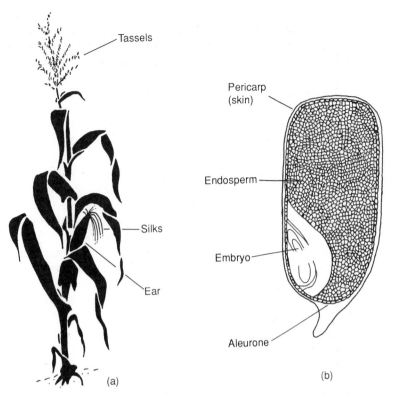

Fig. 8.2 (a) Adult maize plant showing its reproductive structures. (b) Drawing of a maize kernel showing the embryo and its surrounding endosperm and aleurone tissue.

the same plant. After pollination a single ear, therefore, produces many *kernels*. Each kernel is made up of an embryo surrounded by nutritive tissue called *endosperm*, as well as an *aleurone* layer of cells and a pericarp (skin) (Fig. 8.2b). The kernels have several characteristics which can be used for genetical studies, including the quality and colour of the starch in the endosperm and the colour of the surrounding aleurone layer*.

On ears produced on F_1 plants, the kernels, therefore, comprise a population of F_2 progeny, which can be scored for kernel phenotype. This makes maize ears a very convenient way of studying genetic ratios.

Kernel form

Kernels are either swollen with a *smooth* outline, or collapsed with a *wrinkled* outline (Fig. 8.3b). These features are equivalent to the round and wrinkled phenotypes of Mendel's peas. The difference between the two is determined largely by the amount of sugar versus starch in the mature endosperm. *Sugary* endosperm dries out readily, causing the kernels to collapse and become wrinkled, much more so than *starchy* endosperm, the kernels of which thus remain swollen and smooth in outline.

High concentrations of sugar in kernels are produced by two genes, whose symbols, names and linkage positions are:

---

* These nutritive tissues are, in fact, *triploid*, though for the purposes of the present exercises this fact can be ignored.

sh2 :   shrunken endosperm, 3 – 111.2 (= chromosome 3, map position 111.2) and

su1 :   sugary endosperm, 4 – 71.

Their corresponding dominant alleles are **Sh2** and **Su1**, and these produce *starch* in the endosperm.

Kernels that are homozygous for *either* of the recessive alleles contain sugar, dry out at maturity and thus are collapsed and wrinkled. If both dominant alleles are present (each in either homozygous or heterozygous condition) the kernels contain starch and thus are swollen at maturity, with smooth outlines. The genes are probably involved in the conversion of sugar to starch via enzyme action: dominant alleles (**Sh2** and **Su1**) produce enzymes that convert sugar to starch, while the recessive alleles (**sh2** and **su1**) produce no enzyme or defective enzymes and thus sugar is not converted to starch. In the heterozygous condition (**Sh2 /sh2** or **Su1/su1**) the quantity of enzyme produced by the dominant alleles is sufficient to produce starch in the quantity necessary to keep the kernels in a swollen, smooth state.

In this chapter we will refer to the phenotypes as *smooth* and *wrinkled*, since this is their most obvious characteristic. In some literature the phenotypes are referred to as *starchy* and *sweet*; or *shrunken* when the **sh** gene is involved.

In symbolizing the crosses later we will ignore for the sake of convenience the numbers following the two genes concerned. These numbers merely indicate the existence of other genes having similar effects.

Kernel colour

The dark (red or purple) colours of maize kernels result from the production of an anthocyanin pigment in the aleurone tissues. If this pigment is present the kernels are said to be '*coloured*' (red or purple), while if it is absent the kernels are said to be '*colourless*' (white or yellow) (Fig. 8.3b).

The basic colour of the anthocyanin pigment is *red*. The main genes involved in its formation and of interest here are listed below, where symbols, names and map positions are given.

a : anthocyanin-less, 3 – 111
c : colourless aleurone, 9 – 26 and
r : colourless aleurone and plant, 10 – 57

For each of these recessive alleles there is a corresponding dominant allele (**A**, **C** and **R**, respectively). A kernel has to carry at least one of each of these three dominant genes for red pigment to be produced and thus for the kernels to be 'coloured'. Homozygosity for *any* of the recessive alleles (i.e. **a/a** or **c/c** and **r/r**) results in absence of pigment and thus 'colourless' kernels.

Two additional genes affecting anthocyanin pigment production are as follows.

Pr : Purple, 5 – 46. The presence of this dominant gene changes the colour of the anthocyanin pigment from red to purple. The recessive allele (**pr**), when homozygous, has no effect. Thus purple kernels have the genotype **Pr/-**, while red ones have the genotype **pr/pr**.

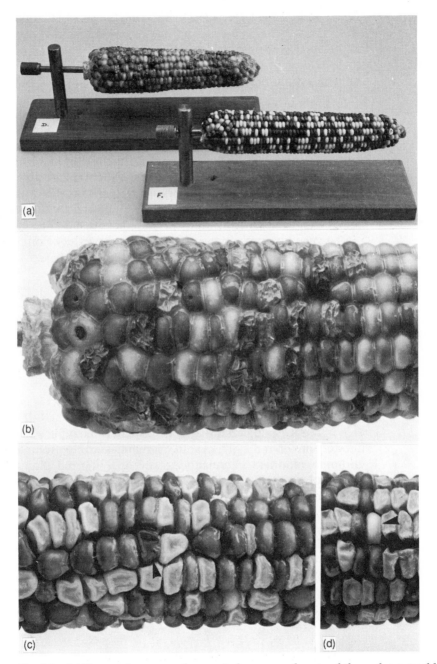

Fig. 8.3 (a) Two maize ears each mounted on a wooden stand through a rotatable spike for ease of scoring. (b) Portion of a maize ear segregating 9 : 3 : 3 : 1 for smooth vs wrinkled and coloured vs colourless. Colour is accentuated in wrinkled kernels because of pigment concentration. The kernel at centre left is marked by a large dot to indicate where counting should start and finish. Two weevil holes are evident! (c) Separate portions of a maize ear (H of Table 8.2) showing parental (red, smooth and yellow,wrinkled) and recombinant (red, wrinkled and yellow, smooth) kernels. The two recombinant kernels are labelled with tailless arrows.

I : Inhibitor, 9 – 26 (very close to but distinct from **C**, though it is sometimes symbolized **C-I**). This is a *dominant inhibitor* of red (and therefore purple) pigment. The recessive allele (i) has no inhibitory effect. Thus red or purple kernels have the genotype i/i, while some kernels are colourless because they carry the **I** gene, in homozygous or heterozygous condition.

In addition to anthocyanin pigment, maize kernels also produce yellow carotenoid pigment *in their endosperm*. A gene affecting production of this pigment is:

Y : Yellow endosperm, 6–17. Presence of this dominant allele in a kernel produces a yellow endosperm. Homozygosity for the recessive allele (y) results in yellow pigment not being produced, and thus white kernels. The presence of **Y** or **y** is masked by red or purple anthocyanin pigment. As noted above, white or yellow kernels are referred to as 'colourless' (though strictly they are colourless only in relation to the absence of anthocyanin pigment).

The genetic compositions of kernels in respect of colour are as follows, where the - indicates either the dominant or the recessive allele.

Coloured
    Red : A/-, C/-, R/-, pr/pr, i/i
    Purple : A/-, C/-, R/-, Pr/-, i/i
Colourless
    Yellow : Y/- and either I/- or a/a or c/c or r/r
    White : as for yellow, but y/y rather than Y/-.

A scheme depicting how these colour genes may act is given in Fig. 8.4, where biochemical substrate, intermediates and enzymes are all given symbols. From this figure it can be seen that gene control of enzyme synthesis will explain the production of red anthocyanin pigment, starting from an initial substrate X through colourless intermediates; and how the red pigment is converted to a purple pigment. If a **pr** allele is present in homozygous condition then no enzyme is produced for this step in the sequence and thus only red pigment will be produced; while if any one of the other three alleles (**a**, **c** or **r**) is present in homozygous condition, neither red nor purple anthocyanin pigment is produced.

The dominant inhibitor gene (**I**) is thought to produce a substance which inhibits one of the first three steps in the above process, perhaps by inhibiting the action of its nearby **C** gene or the enzyme produced by this gene.

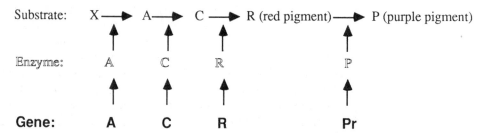

Fig. 8.4 Scheme depicting the action of aleurone colour genes **A**, **C**, **R** and **Pr** of maize during the formation of red and purple anthocyanin pigments, by enzymatic action on substrate (symbolized X) and intermediate compounds.

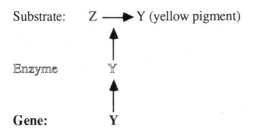

Substrate: Z ──▶ Y (yellow pigment)

Enzyme   Y

Gene: Y

Fig. 8.5 Scheme depicting the action of the endosperm colour gene Y of maize during the formation of carotenoid pigment via enzymatic action on substrate Z.

Finally, the **Y** gene is considered to act in a similar way to produce carotenoid pigment for endosperm colouration (Fig. 8.5).

Summary
Since the colour genes and the **su** gene are in separate chromosomes they segregate independently of each other during meiosis and thus give typical dihybrid ratios (9 : 3 : 3 : 1 and 1 : 1 : 1 : 1) in the F_2. Moreover, the colour genes amongst themselves are also unlinked. Because they affect the same phenotypic character (kernel colour) their dihybrid (and trihybrid) ratios are modified in various ways to give 9 : 7, 13 : 3 and other ratios. Study of these ratios, especially their biochemical interpretation, gives a valuable insight into how genes interact to produce unexpected patterns of inheritance.

Finally, the **Sh** and **A** genes are closely linked in chromosome 3 and therefore can be used in a simple linkage exercise.

Thus these maize ears illustrate a wide range of genetic phenomena, from the rules of heredity using simple monohybrid and dihybrid ratios to modified ratios brought about by interacting genes (epistasis), as well as linkage and crossing over.

PROCEDURES
For Seedlings

Different species sometimes have specific conditions for seed germination, which are usually provided as supplementary information with the purchase of the seed. Maize (corn) and sorghum are easy to produce seedlings from, making them especially good to use in a genetics laboratory. Because of their smaller size, tomato and especially tobacco seed need more careful handling. Generally seedlings are best produced in shallow plastic or wooden trays (Fig. 8.1d) in a seed mix (peat and pumice or equivalent), which should be sterilized beforehand. Seed mix purchased commercially contains sufficient fertilizer for our purposes, so that further application of fertilizer is not necessary, unless seedlings are to be grown on so as to maintain segregating lines (in which case they will need planting out into individual pots). Before planting seed, the mix should be well moistened and firmed down in the tray. Seeds should be planted no less than 5 cm apart in *regular* rows and tiers, so that subsequent scoring of seedlings is easy. Cover the seed with about 1 cm of mix, firm down and cover the tray with a sheet of glass or plastic to prevent drying. Label the tray and leave it in a warm place in dim light. Seedlings will normally appear within 5–10 days, depending on the species and the conditions of growth. Remove the cover of the tray when seedlings appear and place the tray in good light. Seedlings will be ready to score in a further 1 – 2 weeks, or less in some cases. At least 100 seedlings should be produced to allow for statistical testing of the results.

The purchase of fresh seed each year is recommended. This is because a high germination rate with even appearance of seedlings greatly helps in obtaining quick, accurate data. However, segregating lines can be maintained by self-pollinating appropriate plants, thus avoiding the need to purchase seed each year. This approach, however, is very time consuming and not recommended unless purchase of fresh seed each year is difficult.

For Maize Ears

A number of biological supply houses offer maize ears showing a variety of standard and modified genetic ratios involving the genes described in the Introduction and Background section. Usually the ears have already been fumigated to avoid importation problems. They should be lacquered to improve appearance and permanency, stored carefully so as not to dislodge kernels, and kept clear of vermin such as weevils and mice! Mounting the ears on a rotatable spike on a wooden stand (Fig. 8.3a) provides a convenient way of labelling and handling them and for permanently marking a start point for counting.

EXERCISES

1 With Seedlings

Approach

Two alternative approaches to using seedlings in a genetics class are possible. They can be used as *demonstrations* of Mendelian ratios in plants, in which case trays of seedlings are put on display along with an explanation of how they were obtained and what they illustrate. For a more interesting and challenging exercise, however, trays of seedlings can be presented as 'unknowns', to which students attempt to give solutions.

Especially for the latter approach students should proceed as follows:

1 Decide on the contrasting pair or pairs of characters shown.
2 Score the seedlings, using previously prepared tally sheets.
3 Derive and test a hypothesis concerning the pattern of inheritance that the data illustrate.
4 Predict the phenotypes and genotypes of the parents and F_1 of the cross concerned.

Results and explanations

Typical data for some of the main seedling types currently available are given in Table 8.1.

The explanations for these data are given below, where we use the *general* symbols **A** (for a dominant trait and its dominant allele) and **a** (for the corre-

Table 8.1 Seedling data of various plant species obtained from approximately 100 seeds (one packet as supplied by Carolina Biological Supply Company), along with predicted ratios, chi-squared (χ^2) values and types of inheritance. *d.f.* = degrees of freedom, *P* = probability (given to the nearest $1/10$th or $1/20$th; see Appendix Table A.2). For description of characters see p. 152.

Plant	Contrasting characters and numbers scored	Predicted ratio	χ^2	d.f	P	Type of inheritance
Sorghum	72 red, 28 non-red	3 : 1	0.48	1	0.5	
Sorghum	68 green, 25 albino	3 : 1	0.18	1	0.6	Monohybrid
Maize (corn)	55 tall, 12 dwarf	3 : 1	1.80	1	0.2	cross, with
Maize (corn)	61 green, 16 albino	3 : 1	0.73	1	0.4	complete
Tomato	48 cut, 22 potato	3 : 1	1.54	1	0.2	dominance
Tomato	57 hairy, 28 hairless	3 : 1	2.86	1	0.1	
Soybean	24 green, 36 yellow-green, 16 yellow	1 : 2 : 1	1.89	2	0.4	Monohybrid cross, with incomplete dominance
Tomato	18 green, 40 yellow-green, 8 yellow	1 : 2 : 1	6.0	2	0.05	
Maize (corn)	43 tall & green, 11 tall & albino, 22 dwarf & green, 5 dwarf & albino	9 : 3 : 3 : 1	4.35	3	0.2	Dihybrid cross, with complete dominance
Sorghum	54 green, 41 albino	9 : 7	0.02	1	0.9	Dihybrid cross, with epistasis

sponding recessive trait and allele); similarly, where necessary, for **B** and **b**. Also, we write the genotypes in linear form (**A/A**, **A/a**, **a/a** etc.) in which the / indicates the pair of homologous chromosomes in which the appropriate genes are located. Note also that for the genotype **A/A**, **B/B**, for example, the comma separates the two genes located in different pairs of chromosomes; and for the genotype **A/-** the – means either **A** or **a**. See Chapter 3 for further explanation and alternative ways of writing genotypes and crosses.

Monohybrid ratio with complete dominance (3 : 1)
In each of the first six groups of seedlings of Table 8.1 a predicted 3 : 1 ratio is confirmed by chi-squared analysis, in that each probability is greater than the 0.05 level (see Appendix for explanation). Thus any deviation from an exact 3 : 1 ratio is explained solely by chance (sampling error). The 3 : 1 ratio implies that the contrasting characters are determined by the action of a single pair of alleles

in which one allele is completely dominant over the other. This is symbolized as follows.

Parents: A/A   x   a/a   (phenotype A x a)
Gametes: all A   all a
F_1:   All A/a   (phenotype A, because of
                                                    dominance of the A allele)

Segregate the alleles into different cells during meiosis in the F_1 to produce equal numbers of A and a gametes. Self-pollinate the F_1.

F_2 genotypes: A/A   A/a   a/a
Ratio of genotypes: 1   :   2   :   1
Ratio of phenotypes: 3 A   :   1 a

For the green/albino situation (sorghum, maize) this full cross is not possible; and students can be quizzed as to how seed segregating for these traits are produced, given the lethality soon after germination of the a/a genotype. The answer, of course, is by self-pollination of plants shown by progeny tests to be heterozygous for the albino allele.

Monohybrid ratio with incomplete dominance (1 : 2 : 1)
For the two lots of green, yellow-green, yellow seedlings of Table 8.1 the occurrence of three phenotypes and their relative frequencies suggest a 1 : 2 : 1 ratio; and this is confirmed by their chi-squared values. This ratio implies monohybrid inheritance as above, but in which the heterozygote A/a is distinguishable. The F_2 phenotypic ratio is then the same as the F_2 genotypic ratio.

Note that production of seed segregating for the lethal yellow condition does not require progeny testing, because the heterozygotes can be visually identified.

Dihybrid ratio with complete dominance (9 : 3 : 3 : 1)
For the third group of seedlings in Table 8.1 the four phenotypes and their relative frequencies immediately suggest a standard dihybrid ratio of 9 : 3 : 3 : 1, which is confirmed by the chi-squared value. This ratio implies the action of *two* pairs of alleles which segregate independently of each other, and in which one allele of each pair is completely dominant over its counterpart allele.
    The standard dihybrid ratio is explained as follows.

Parents: A/A, B/B   x   a/a, b/b (phenotype A B x a b)
Gametes: all AB   all ab
F_1:   all A/a, B/b   (phenotypes A B because
                                                   of dominance)

Segregate the alleles independently into different cells during meiosis in the F_1 to produce equal numbers (25%) of AB, Ab, aB and ab gametes. Self-pollinate the F_1.

F_2 genotypes:	A/-, B/-	A/-, b/b	a/a, B/-	a/a, b/b
F_2 phenotypes:	AB	Ab	aB	ab
Ratio of phenotypes:	9 :	3 :	3 :	1

This scheme will not explain fully the origin of the dihybrid ratio of Table 8.1, because the albino condition is lethal and thus the parent genotype a/a, b/b does not survive to adulthood. Rather, F_1 seed are produced by crossing homozygous tall plants that are shown by progeny tests to be heterozygous for the albino trait with homozygous dwarf plants also shown to be heterozygous for the albino trait. Thus if the albino allele is designated b the cross to set up is as follows.

Parents	A/A, B/b	x	a/a, B/b (phenotype: tall green x dwarf green)
Gametes:	$\frac{1}{2}$ AB, $\frac{1}{2}$ Ab	$\frac{1}{2}$ aB, $\frac{1}{2}$ ab	
F_1:	$\frac{1}{4}$ A/a, B/B	$\frac{1}{2}$ A/a, B/b	$\frac{1}{4}$ A/a, b/b

Of these F_1 the latter will die because of their albino condition; while the first will *not* produce albino progeny when self-pollinated, and are discarded. Those surviving F_1 that produce albino progeny when self-pollinated are used to obtain further seed that will give the desired $9 : 3 : 3 : 1$ ratio.

Dihybrid ratio with epistasis (9 : 7)

In the last group of seedlings of Table 8.1 the 54 green : 41 albino seedlings cannot represent a 1 : 1 testcross ratio. This is because for a lethal gene a testcross (A/a x a/a) is impossible, since the genotype a/a does not survive to maturity.

Since monohybrid inheritance cannot easily explain this ratio, the next step is to consider modified dihybrid inheritance. 54 is close to $\frac{9}{16}$ths of the total (95) and 41 close to $\frac{7}{16}$ths. 9 : 7 is a modified dihybrid ratio in which *two* distinct genes each separately produces the albino condition. Thus in the standard dihybrid ratio illustrated above, both a *and* b represent albino alleles. In the F_2 only the genotype A/-, B/- gives green seedlings, while all three other genotypes give albino seedlings, because of homozygosity for one or other or both albino alleles. Thus a 9 : 7 ratio of green : albino seedlings is produced.

Because we are dealing with a lethal (albino) condition, seed segregating 9 : 7 for green and albino seedlings are, in fact, obtained from crossing plants known to be heterozygous each for a different albino allele. The seed derived by self-pollination of the progeny from this cross are then tested so as to select lines that give the desired 9 : 7 ratio of green : albino. The procedure is given below, where a and b are the two alleles for the albino condition (and A and B their respective green alleles).

Parents:	A/A, B/b		x	A/a, B/B	
Gametes:	$\frac{1}{2}$ AB, $\frac{1}{2}$ Ab			$\frac{1}{2}$ AB, $\frac{1}{2}$ aB	
Progeny:	$\frac{1}{4}$ A/A, B/B	$\frac{1}{4}$ A/A, B/b	$\frac{1}{4}$ A/a, B/B	$\frac{1}{4}$ A/a, B/b	

Of these progeny the first, when self-pollinated, will produce seedlings none of which will be albino; while the second and third produce seedlings of which only

$1/4$ will be albino. Only the last genotype (**A/a, B/b**) will, when self-pollinated, produce the desired ratio of 9 green : 7 albino.

These seedlings, therefore, provide a useful demonstration of how two genes that affect the same phenotypic character can interact to give an unexpected pattern of inheritance. They are best given to students as an 'unknown', to present them with an interesting challenge to explain a ratio of green : albino that cannot be 1 : 1, to see what sort of an explanation they can come up with.

2 With Maize Ears

Approach

This exercise is best carried out by presenting appropriate maize ears to students as 'unknowns', for which solutions are attempted. The solutions to some ears are more difficult than others, so choice should be exercised according to the desired level of teaching. Of the ears for which data are given in Table 8.2, A – E are straightforward and thus suitable for elementary genetics classes. F and G are appropriate for those who have been introduced to epistasis, and H for those studying linkage and crossing over.

The ears are best labelled so as not to disclose their type. Knowing beforehand the actual numbers of each kernel type (not simply their ratio) provides an imme-diate check on the accuracy of student counting. Provide students with background knowledge sufficient for them to interpret the ratios they will obtain. If counting becomes tedious the work should be shared amongst students, or for some ears students can be given data obtained previously. Indeed, this whole exercise can be run as demonstrations of appropriate ears followed by analysis of given data (such as those of Table 8.2).

For analysing the ears proceed as follows.

1 Determine the contrasting characters shown by each ear. This should be done by looking at form and then colour. For form, the contrasting characters should be designated *smooth* (=starchy) and *wrinkled*. Note that the small depression on the top of some kernels should be ignored and not confused with the wrinkled phenotype. For colour, the contrasting characters should be designated *coloured* (and then, if coloured, whether they are red or purple) or *colourless* (and then white or yellow). Note that wrinkled kernels are darker than equivalent smooth ones, because of pigment concentration.

2 Count the kernels of each type. This is done by using a tally sheet or hand tally-counter. Start at a recognizable point and proceed systematically along each row. Use the *back* of a pencil or biro for counting, so as not to mark the kernels.

3 Construct a hypothesis to explain the genetic ratio given by the ear. Where the ratio is not obvious a guide can be obtained by dividing all the totals by the smallest one, and then either rounding off to the nearest whole or multiplying throughout by 2, 3, etc. until whole numbers are approximated. Alternatively, a ratio may be suggested by prior knowledge of the genetics of kernel colour, or arrived at by intuition.

4 Test the hypothesis using chi-squared (see Appendix). Where two or more ratios are possible (e.g. 1 : 1, 9 : 7) all possibilities should be tested and then a judgement made accordingly.

5 Determine the genotype and phenotype of the original parents and F_1; and then illustrate the cross. Always assume the parents were homozygous (true breeding) for the genes (traits) concerned. It is sensible to show the *full* genotype of the parents initially, but subsequently to consider only those genes which are *different* when showing the details of the cross. For example, true breeding purple crossed with true breeding red can be given initially as **R/R, A/A, C/C, Pr/Pr, i/i** x **R/R, A/A, C/C, pr/pr, i/i** but subsequently need only be shown as **Pr/Pr** x **pr/pr**, since only the purple pigment alleles vary. Ear B of Table 8.2 is explained in detail to illustrate fully how the solutions are arrived at.

6 Provide a biochemical–genetic interpretation of the situation.

Results and explanations

Actual data from selected ears are shown in Table 8.2 and are explained as follows.

Table 8.2 Numbers of kernel types (phenotypes) in eight different maize ears. Under Ear, the number of ears of each type scored is given in brackets.

Ear	Smooth	Wrinkled	Coloured (purple)	Coloured (red)	Colourless (yellow)	Colourless (white)	Coloured (red), smooth	Coloured (red), wrinkled	Colourless (yellow), smooth	Colourless (yellow), wrinkled	Total
A(1)	501	182									683
B(1)			439			165					604
C(1)			283		261						544
D(1)							340	119	90	37	586
E(2)							285	264	270	243	1062
F(2)			499		403						902
G(1)				92	414						506
H(3)							1020	1	2	1109	2132

Monohybrid ratio (3 : 1)
Ear A. 501 smooth, 182 wrinkled
Hypothesis
Monohybrid ratio (3 : 1).
$\chi^2 = 0.99$, *d.f.* = 1, $P = 0.3$. The hypothesis is accepted.

Explanation

Parents:	**Su/Su** x **su/su** (or **Sh/Sh** x **sh/sh**). Smooth x wrinkled.		
F_1:	**Su/su** All smooth (smooth,**Su**, is dominant to wrinkled,**su**).		
	Self-pollinate the F_1 (**Su/su** x **Su/su**).		
F_2:	**Su/Su**	**Su/su**	**su/su**
	(smooth)	(smooth)	(wrinkled)
	1 :	2 :	1
		3 smooth :	1 wrinkled

Biochemical interpretation
The **su** allele prevents the conversion of sugar to starch and thus causes the endosperm to shrink and become wrinkled (p.156). In the heterozygote (**Su/su**) the quantity of functional enzyme produced by the one **Su** allele is sufficient to form starch in the quantity necessary to keep the kernels in a swollen (smooth) state.

Ear B. 439 purple, 165 white
Hypothesis
Monohybrid ratio (3 : 1).
$\chi^2 = 1.73$, *d.f.* = 1, $P = 0.2$. The hypothesis is accepted.
Explanation
The 3 : 1 ratio implies a single pair of contrasting characters determined by a single pair of alleles, in which one allele is completely dominant over the other. All kernels must therefore be homozygous for **y** and homozygous for **Pr**, for otherwise yellow and red progeny would have appeared in the F_2. Hence the ear derives from a cross coloured (purple) x colourless (white). The full genotypes of the parents were thus **A/A,C/C,R/R,Pr/Pr,i/i,y/y** (purple aleurone masking white endosperm) and **a/a,C/C,R/R,Pr/Pr,i/i,y/y** or equivalent with **c/c** or **r/r** instead of **a/a** (colourless aleurone, white endosperm and thus overall white). In symbolizing the cross we ignore all genes except those for which the parents differ i.e. **A/A** and **a/a**.

Parents:	**A/A** x **a/a** (or equivalent for **C, c** or **R, r** alleles).		
	Coloured (purple) x colourless (white).		
F_1:	All **A/a** (coloured, **A**, dominant to colourless, **a**).		
	Self-pollinate the F_1 (**A/a** x **A/a**).		
F_2 :	**A/A**	**A/a**	**a/a**
	(coloured)	(coloured)	(colourless)
	1 :	2 :	1
		3 coloured (purple) :	1 colourless (white)

Biochemical interpretation
A defect is present in the biochemical pathway between initial substrate and formation of red pigment due to the presence of **a/a** (or **c/c** or **r/r**); the pathway from red pigment to purple pigment is intact (p.158). Purple masks the absence of yellow pigment in the endosperm.

Monohybrid testcross ratio (1 : 1)
Ear C. 283 purple, 261 yellow
Hypothesis
Monohybrid testcross ratio (1 : 1).
$\chi^2 = 0.89$, $d.f. = 1$, $P = 0.4$. Hypothesis accepted.

Explanation

Parents and F_1:	as for ear B, except that each was Y/Y rather than y/y; purple x yellow.
	Testcross the F_1 (**A/a x a/a**).

F_2 :	**A/a**	**a/a**
	coloured (purple)	colourless (yellow)
	1	:	1

Biochemical interpretation
As for ear B, except yellow pigment is produced in the endosperm.

Dihybrid ratio (9 : 3 : 3 : 1)
Ear D. 340 red smooth, 119 red wrinkled, 90 yellow smooth, 37 yellow wrinkled
Hypothesis
Dihybrid ratio (9 : 3 : 3 : 1).
$\chi^2 = 4.7$, $d.f. = 3$, $P = 0.2$. Hypothesis accepted.

Explanation
This is a combination of the crosses for ears A and B, except that the parents were **Y/Y** and thus had yellow rather than white endosperm; and were **pr/pr** so that the coloured phenotype is red rather that purple. Since the genes concerned are in separate chromosomes they segregate independently during meiosis of the F_1, and thus produce a 9 : 3 : 3 : 1 ratio in the F_2.

Parents :	**A/A, Su/Su x a/a, su/su**
	(or the equivalent with **C,c** or **R,r** and **Sh,sh** alleles).
	Coloured (red), smooth x colourless (yellow), wrinkled.
	Alternatively, the cross might have been **A/A, su/su x a/a, Su/Su**; red,wrinkled x yellow, smooth. For genes on different chromosomes the F_1 and F_2 are the same for these two arrangements of the dominant and recessive traits.
	Both parents were also **pr/pr** and **Y/Y.**
F_1 :	All **A/a , Su/su** (red, smooth, because of dominance of **A** and **Su**). Self-pollinate the F_1.

F_2 :	**A/-, Su/-**	**A/-, su/su**	**a/a , Su/-**	**a/a , su/su**
	(red, smooth)	(red,wrinkled)	(yellow, smooth)	(yellow,wrinkled)
	9	: 3	: 3	: 1

Dihybrid testcross ratio (1 : 1 : 1 : 1)
Ear E. 285 red smooth, 264 red wrinkled, 270 yellow smooth, 243 yellow wrinkled
Hypothesis
Dihybrid testcross ratio (1 : 1 : 1 : 1).
$\chi^2 = 3.4$, *d.f.* = 3, *P* = 0.3. Hypothesis accepted.

Explanation
Parents and
F_1:   as for ear D.
              Testcross the F_1 (A/a, Su/su x a/a, su/su).
F_2 :   as for ear D except that the four phenotypes are produced in a
              1 : 1 : 1 : 1 ratio.

Dihybrid ratios with epistasis (9 : 7 and 13 : 3)
Ear F. 499 purple, 403 yellow
Hypothesis 1
Monohybrid testcross ratio (1 : 1), as for ear C.
$\chi^2 = 10.2$, *d.f.* = 1, *P*<< 0.01. The hypothesis is *rejected*; this is *not* a good fit to a 1 : 1 ratio.
We therefore consider a dihybrid situation in which *two* genes are interacting to produce a modified 9 : 3 : 3 : 1 ratio. 499 is close to $\frac{9}{16}$ths of the total (902) and 403 close to $\frac{7}{16}$ths. 9 : 7 is a modified dihybrid ratio in which the last three progeny types of the standard dihybrid ratio are combined and produce yellow kernels. Note that this solution requires some prior knowledge of the way(s) in which a dihybrid ratio can be modified when two genes affect the same phenotypic character.
Hypothesis 2
A modified dihybrid ratio (9 : 7).
$\chi^2 = 0.3$, *d.f.* = 1, *P* = 0.6. Hypothesis accepted.

Explanation
Parents :   A/A, C/C x a/a, c/c (or equivalents, as in ear D).
              Parents were also **Pr/Pr** and **Y/Y**.
              Coloured (purple) x colourless (yellow).

F_1:   All **A/a, C/c** (coloured,purple).
              Self-pollinate the F_1.
F_2:   A/-, C/-   a/a, C/-   A/-, c/c   a/a, c/c
              Coloured (purple) (yellow)   (yellow)   (yellow)
              9   :   3   :   3   :   1

              9 coloured (purple)   :   7 colourless (yellow)

Biochemical interpretation
The cross involved *two* genes producing separate metabolic blocks to the synthesis of red pigment. The F_2 yellow progeny differ according to whether they carry one or other or both of these metabolic blocks.

Ear G. 414 yellow, 92 red
Hypothesis
This is clearly not a 3 : 1 ratio. We therefore suspect a modified dihybrid ratio. 414 is close to $^{13}/_{16}$ths of the total (506) and 92 close to $^{3}/_{16}$ths. 13 : 3 is a modified dihybrid ratio obtained by combining the first, third and fourth categories of the standard 9 : 3 : 3 : 1 ratio.
$\chi^2 = 0.1$, *d.f.* = 1, *P* = 0.75. Hypothesis accepted.

Explanation
For the first category of the standard dihybrid ratio to produce *yellow* kernels, one of the two dominant alleles must *inhibit* anthocyanin pigment formation. Hence the cross must have involved the dominant inhibitor gene I, in addition to one other gene affecting colour.

Parents :   i/i, R/R x I/I, r/r (or equivalents as in ear D).
               Coloured (red) x colourless (yellow).
               Both parents were also **pr/pr**.
F_1:   All **I/i, R/r** (yellow, note!).
               Self-pollinate the F_1.

F_2:

I/-, R/-	i/i, R/-	I/-, r/r	i/i, r/r
yellow	red	yellow	yellow
9 :	3 :	3 :	1
	13 yellow :	3 red	

Biochemical interpretation
The product of the I gene inhibits formation of anthocyanin pigment (p.158). Thus three genotypically distinct categories of yellow kernels are produced in the F_2.

Linkage
Ear H. 1020 red smooth, 1 red wrinkled, 2 yellow smooth, 1109 yellow wrinkled
Hypothesis
This is clearly not a standard dihybrid ratio. Note that two categories are much more frequent than the other two. This is typical of a *dihybrid cross involving linkage*. The categories with the large numbers represent non-recombinant (parental) F_2 progeny, while the smaller categories represent the recombinant (crossover) progeny.

Explanation
The only two *linked* genes affecting kernel colour and kernel form are **A** and **Sh**.

Parents: A Sh   x   a sh   (red, smooth x yellow, wrinkled).
              ────   ────
              A Sh   a sh

F_1 All   A Sh   (red, smooth).
                        ────
                        a sh

Testcross the F_1

i.e. $\dfrac{\text{A Sh}}{\text{a sh}}$ x $\dfrac{\text{a sh}}{\text{a sh}}$

Progeny:

$\dfrac{\text{A sh}}{\text{a sh}}$	red, smooth	Non-recombinant type
$\dfrac{\text{a sh}}{\text{a sh}}$	yellow, wrinkled	Non-recombinant type
$\dfrac{\text{A sh}}{\text{a sh}}$	red, wrinkled	Recombinant type
$\dfrac{\text{a Sh}}{\text{a sh}}$	yellow, smooth	Recombinant type

Comments

The cross is similar to that of ear E, except that linkage between **A** and **Sh** is involved, rather than independent segregation between **A** and **Su**. Ears E and H, therefore, provide a superb demonstration of the contrasting effects of genes being in different chromosomes versus being linked in the same chromosome.

The linkage value is given as the number of recombinants divided by the total number of F_2 progeny (recombinants plus non-recombinants) – i.e. 3/2132, = 0.14% (see Chapter 3 for further information). The published map positions for **A** and **Sh** are 111 and 111.2 (p. 156), meaning that the loci are 0.2 map units apart (very close together!). This gives an expected recombination frequency of 0.2%, to which the observed value (0.14%) is very close. Recombinant and non-recombinant kernels are illustrated in Fig. 8.3d.

Other ear types are available from supply houses, including those illustrating a dihybrid ratio modified by epistasis to give 9 purple : 3 red : 4 yellow; and a *trihybrid* ratio modified by epistasis to give 27 red : 37 yellow.

These are recommended for advanced genetics classes. Their explanations follow easily from what has been given above.

REFERENCES

Sprague G.F. (Ed.) (1955). *Corn and Corn Improvement*. London, New York, Academic Press.
Srb A.M., Owen R.D. and Edgar R.S. (1965). *General Genetics*. Second edition. San Francisco, W.H. Freeman & Co.

Chapter 9 GENETICS WITH HUMANS

Human genetics can be a sensitive area of study, and also one which involves small added risks where procedures involve taking blood samples. In some countries, blood sampling of students is prohibited in educational establishments. In other countries an unlicensed person can sample only his or her own blood. Local regulations on this matter, therefore, should be carefully checked and strictly followed. Whatever the local circumstances, unqualified taking of quantities of blood other than those involved in a finger prick is dangerous. It is essential, therefore, to obtain qualified medical advice before a programme on making slides of human chromosome, for example, is embarked upon.

Exploring family relationships via blood group testing, for example, is inadvisable in a class situation, because of the possible damaging effects of exposed infidelities. There is also a small risk of exposing an abnormal chromosome constitution if prepared slides are made from students' own blood. Information gained from blood groups and chromosome karyotypes in a class situation should never be regarded as authentic substitutes for medically determined information.

Because of these considerations, some of the exercises in this chapter are most appropriate for undergraduate university or polytechnic classes in genetics. Others, however, are appropriate or can be modified easily for use in senior school classes.

INTRODUCTION AND BACKGROUND

Students are highly motivated by the application of genetics to humans, and it is rewarding to include in a genetics laboratory course at least one exercise involving our own species. Humans, however, are by no means an easy subject of study when it comes to genetics. We cannot experiment with humans in the way that we can with other organisms such as Drosophila and maize, but there are some exercises involving common human traits, human blood groups and human chromosomes that are amenable to study, and these form the subject of this chapter.

Genetics of some Common Traits

A number of common human traits are amenable to study in a genetics laboratory (Table 9.1). Three of these traits form the basis of Exercise 1 in this chapter, namely PTC tasting, crown hair whorl and tongue rolling.

Table 9.1 Alternative forms of common human traits known to have a genetic basis. See McKusick (1986) for further examples and genetic information.

Trait	Alternative forms
Colour vision	Normal colour vision vs reduced or no sensitivity to red light or green light
PTC tasting	Can or cannot taste weak solutions of phenylthiocarbamide (see text)
Arm folding	Right vs left arm on top when arms are folded
Hand clasping	Fingers of right vs left hand lie above corresponding fingers of other hand when clasped
Handedness	Predominantly right vs predominantly left handed
Hair colour	Naturally black, brown, blond, etc.
Hair whorl	Crown hair whorl turns clockwise or counter-clockwise (see text)
Hair curliness	Naturally curly, straight or wavy
Tongue rolling	Can or cannot roll tongue lengthwise (see text)
Tongue folding	Can or cannot fold tongue backwards from tip
Eye colour	Brown, blue, grey, etc.
Ear lobe attachment	Free vs 'attached' (=no) earlobes
Finger length	Index (first) finger shorter vs longer than ring finger
Finger nail lacuna	Large vs very small or no 'moons' on finger nails
Mid-digital hair	Hairs present or absent on one or more middle digits of fingers
Toe length	Second toe longer vs shorter than first ('big') toe

PTC tasting

Humans vary greatly in their taste reaction to a wide variety of substances, including even salt and sugar. In this respect each of us is biochemically unique, giving us different preferences for the food we eat.

Family studies suggest a strong hereditary basis to human variation in taste ability. A clear example of this is seen in our ability or otherwise to taste weak solutions of the substance phenylthiocarbamide (PTC, also known as phenylthiourea). To some the substance is strikingly bitter, while others cannot taste the substance at all. Only a few people cannot be classified easily as 'taster' or 'non-taster', and there is only a slight difference in taste sensitivity between males and females and amongst people of different ages. The frequency of tasters differs across various racial groups, being almost 100% in American Indians but only about 70% in Caucasians. A number of similar substances such as thiouracil evoke similar taste responses. All have the chemical grouping $NC = S$, which is believed to be responsible for the taste reaction. PTC itself is found naturally in some plants grazed by cattle and so is a normal constituent of human food, in which, however, its concentration is very low and undetectable.

Analysis of family data strongly suggests that PTC taste ability is governed largely by a single gene with a pair of alleles in which a tasting allele (**T**) is dominant to the non-tasting allele (**t**). The heterozygous condition cannot be distinguished, so that tasters are of genotypes **TT** and **Tt**, while non-tasters are **tt**.

Tasting ability is easily tested for, simply by applying a small quantity of the substance via a cotton bud or pipette to the tongue: the response of tasters is immediate.

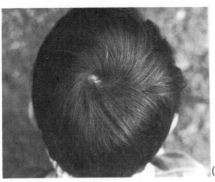 

Fig. 9.1 (a) 'Counter-clockwise' rotation of the crown hair whorl. In 'clockwise' the hair whorl rotates the other way. (b) The 'roller' phenotype, as viewed from the front, in which the sides of the tongue have been folded upwards; in an untrained (7 year old) subject. 'Non-rollers' are unable to fold the tongue in this fashion.

Crown hair whorl

Many morphological traits in humans have a very strong genetic basis to them, in which cultural, learning and other environmental factors play little or no role in their determination. One such trait is crown hair whorl, which rotates in a clockwise direction in some individuals but in a counter-clockwise direction in others (Fig. 9.1a). The clockwise allele (say **C**) is dominant to counter-clockwise (**c**). Occasional individuals have two or even three crown whorls, but these can be disregarded for our purposes.

Tongue rolling

Humans also differ in a wide range of behavioural traits. Some of these have a significant genetic component to their determination, but in which learning, cultural and other environmental factors also play an important role. Included in this category is the ability or otherwise of rolling up the edges of the tongue lengthwise into a U shape (Fig. 9.1b). The 'roller' phenotype may be regarded as the dominant trait (genotypes **RR** and **Rr**), with 'non-rollers' being **rr**. Learning plays some part in deciding which phenotypic class an individual belongs to.

Each of these three traits can be considered for present purposes, therefore, to exist in only the two alternative forms given, and to be determined by a single gene with two alleles, one of which is dominant. This probably simplifies a little the true situation. We can discount possible complications, however, because the purpose of the present study is to introduce students to the types of human traits that are inherited and to the range of genetic variability present in human populations; and to use the relative frequencies of their two forms as *models* in calculating gene frequencies.

ABO Blood Groups

The genetics of human blood groups has been intensively studied and we now have a good understanding of the ABO, MN and Rhesus systems.

In this chapter we will consider only the ABO system, which is the best known and easiest to handle for present purposes. It provides a wealth of data for genetic analysis, demonstrates the phenomenon of multiple allelism (*many* different forms of a gene) and enables gene frequencies in a population to be estimated and compared with others.

The A, B and O blood group phenotypes are determined by a single gene (symbol I for 'isohaemagglutinin') with *three* alleles. There are two codominant alleles, I^A and I^B , and one recessive allele, i. An individual carries only two alleles, of course, so that within the population there are six possible genotypes and, because of dominance effects, four phenotypes. Thus genotypes I^AI^A and I^Ai give blood group A, I^BI^B and I^Bi gives blood group B, I^AI^B group AB and ii group O.

The blood group gene works by determining the form of an *antigenic substance* (mucopolysaccharide) which is secreted onto the surface of the blood cells. The I^A and I^B alleles produce slightly different forms of the antigen. The i allele, on the other hand, produces *no* antigen; hence the basis of the dominance effect. The heterozygote I^AI^B produces both forms of the antigen; hence the basis of the codominance of these two alleles.

The other interesting thing about this blood group system is that a person of one blood type carries in his or her blood *antibodies* against the antigen of a person of a different group. Thus a person of blood group A carries *anti-B* antibodies, and a person of group B carries *anti-A* antibodies. Blood group O individuals have both *anti-A* and *anti-B*, while group AB individuals have neither antibodies, for obvious reasons. For this reason it is not possible to mix the blood from different people in a random way because the incoming red cells, in a transfusion, can be agglutinated (clumped) by the antibodies of the recipient serum. As a rule a person is given blood by transfusion only from an individual of the same blood type.

Table 9.2 summarizes the relevant information for the ABO blood group system.

Table 9.2 Genotypes, antigens and antibodies of the four ABO blood group phenotypes of humans.

Genotype	Blood group phenotype	Antigens	Antibodies
I^AI^A or I^Ai	A	A	anti-B
I^BI^B or I^Bi	B	B	anti-A
I^AI^B	AB	A and B	None
ii	O	None	anti-A and anti-B

Human Chromosomes

Until the early 1950s there was a great deal of controversy about the human chromosome complement, though 48 was the generally accepted chromosome number. Several technical innovations made during the 1950s finally settled the question when the number was accurately described as 46. These technical innovations included the culturing of human blood cells outside of the body (rather than probing for bone marrow cells), the use of phytohaemagglutinin to clump red blood cells and to stimulate mitosis in white blood cells, the use of colchicine to inhibit spindle formation and thus free the chromosomes within the cell, and the use of hypotonic saline to swell the lymphocytes to facilitate the spreading of the chromosomes on microscope slides.

The chromosome number of our species is 2n = 46. Twenty-two pairs of these chromosomes are known as *autosomes*. The remaining two make up the special pair of *sex chromosomes*. Females have a pair of **X** chromosomes, and are thus designated **XX**, while males have one **X** and one **Y** chromosome, and are thus designated **XY**. The **X** and **Y** chromosomes are different in size and centromere position and are thus referred to as being *heteromorphic*. X-linked genes such as those for haemophilia and red-green colour blindness are found in the **X** chromosome but not in the **Y** chromosome, thus making these two chromosomes also genetically different (i.e. *non-homologous*).

Special staining techniques can be employed to produce delicate bands along human (and other) chromosomes, which allow separate identification of all the 22 pairs of autosomes as well as the two sex chromosomes. Studies using these banding techniques have revealed a wealth of information about the standard chromosome complement of humans, and about the range of chromosome abnormalities that are found in human populations. There are many well known (though uncommon) abnormal phenotypes that are related directly to particular chromosome abnormalities, which can be screened for by culturing fetal cells that have been sloughed off into the amnionic fluid, thus checking the karyotype of an unborn person. The study of human chromosomes, therefore, is an important area of medical genetics, as well as having scientific interest.

PROCEDURES

Collecting Data on Common Traits

For PTC tasting

PTC taste papers, with directions for their use, can be purchased from biological supply houses, and these provide an efficient and easy way of distinguishing tasters and non-tasters. The papers are simply sucked and spat out. Alternatively, make up an appropriate solution of phenyl-thiocarbamide (purchasable from many chemical or biological supply houses), by boiling 0.01g in 100ml of distilled water. Apply a drop of the solution towards the back of the tongue of the test subject using a small clean pipette. Record whether or not a bitter taste is detected.

For hair whorl rotation and tongue rolling

Provide photographs (Fig. 9.1) or drawings of the alternative phenotypes to aid classification.

Determining Blood Group Types

Equipment and materials required

Typing kits for determining ABO blood groups are available from biological supply houses. These kits include sterile lancets for finger pricking, testing agents (antisera), student guides and a teacher manual. Alternatively, lancets and antisera can be purchased from medical supply houses. Where it is not permissible or possible to do blood typing in class the appropriate test can sometimes be arranged to be done at a student health clinic, or by offering to donate blood at a hospital or mobile donation unit.

Two kinds of testing agents are used in ABO blood typing. One (called *anti-A*) contains antibodies against antigens A. The other (*anti-B*) contains antibodies against antigens B. Blood from a test person is mixed separately with each antibody and examined for agglutination (clumping).

Blood that is agglutinated by anti-A but not anti-B is type A;
Blood that is agglutinated by anti-B but not anti-A is type B;
Blood that is agglutinated by both anti-A and anti-B is type AB;
Blood that is agglutinated by neither anti-A nor anti-B is type O.

The blood typing procedure is very simple and takes just a few minutes. It is best done at a 'typing station' set up for the purpose and located in an area free of congestion and other students. A person in charge should guide proceedings at the station, which is equipped as follows (Fig. 9.2a).

70% alcohol and swab.
Autclix or other appropriate device for finger pricking.
Replacement lancets (one per student plus a few extras).
Microscope slides (one per student) and marker pen.
0.9% NaCl in a flask with a Pasteur pipette and bulb or in a dropper bottle.
Anti-A and Anti-B blood typing reagents.
Tooth picks.
Container for safe disposal of swabs, lancets, tooth picks and slides.
Band-aids.

The anti-A and anti-B reagents are now produced from special cells grown in culture, rather than being of human origin, and thus there is absolutely no risk of transmission of human disease with use of these reagents. They can be purchased at moderate cost from Biomedical Supply Houses, normally as 10 ml lots in dropper bottles. One such lot is easily enough for typing a class of 150 students. The reagents keep well in a refrigerator for at least 2 years. For ease of recognition the anti-A reagent is dyed blue, and anti-B dyed yellow.

The device used for finger pricking should preferably be of the 'automatic' (pressure release) type and one that cannot easily be re-loaded with a used lancet. Such a device also can be purchased at modest cost from Biomedical Supply Houses. Replacement lancets come in packets of 200 or so and are sterile and protected.

Method

The testing procedure is as follows.

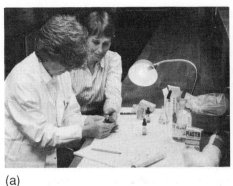 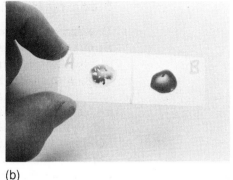

(a)   (b)

Fig. 9.2 (a) 'Blood typing station' showing the equipment needed for ABO blood group typing of students in class. (b) Test slide showing agglutinated (left) and non-agglutinated blood. The subject tested was blood type A.

1 Swab a finger tip with alcohol to clean it. Dispose of the swab in the container provided and allow the finger to dry.
2 Hold the finger to be pricked tightly at its base with the thumb and index finger of the opposite hand, and squeeze upwards from the knuckle towards the finger tip so that the finger tip becomes filled with blood.
3 Prick the finger tip with a newly loaded lancet.
4 Squeeze the finger to produce a droplet of blood. Apply a small drop of blood close to each end of a slide. Label the slide ends 'A' and 'B'.
5 Cover the finger with a plaster if necessary.
6 Add 1 drop of saline (0.9% NaCl) to each drop of blood, followed by one drop of anti-A to the 'A' drop and one drop of anti-B to the 'B' drop. It is best that an assistant adds the anti-A and anti-B, to avoid the reagents becoming contaminated.
7 Mix the contents of each drop with *different* toothpicks. Dispose of the toothpicks.
8 Allow approximately 1 min for agglutination to occur. Examine each area of the slide for agglutination (clumping), ideally by holding it above a white card or with light from a bulb coming from below (Fig. 9.2b). Record the result according to which drop of blood is agglutinated, as follows:

Agglutination in	Blood type
Drop 'A' only	A
Drop 'B' only	B
Both A and B drops	AB
Neither drop A nor drop B	O

It is useful to have a reference slide available for blood type O students, who may query their 'no agglutination' result.

9 Record the results. Dispose of all used materials into a safe container for incineration. Swab down the bench surface with alcohol or disinfectant.

Studying Human Chromosomes

Human chromosomes from prepared slides

Human chromosomes can be studied most easily in a genetics laboratory from prepared slides, purchased from a Biological Supply House. These slides contain many spreads of chromosomes from which quality ones can nearly always be found for setting up as demonstrations. Alternatively, slides may be obtained from a nearby hospital routinely involved in human cytogenetic work.

Making slides of human chromosome

The techniques involved here are more difficult than those used for chromosome studies in plants, for obvious reasons, but are within the ability range of small-sized genetics practical classes. In some situations it may be difficult for students to work through the whole procedure, in which case it will be worthwhile providing the class with ready prepared cells in fixative from which students can then carry on from step 4 in the method that follows. For this purpose one 10 ml vacutainer of blood obtained from a known source will provide ample material for a class of 30 students.

Before embarking on this work the following issues should be considered carefully.

Obtaining blood

Under no circumstances should blood be taken by anyone other than a nurse or doctor. Unqualified taking of blood is dangerous. Most schools and universities have access to nearby health services and one of these is usually more than willing to take the small quantities of blood needed.

Hygiene

For health reasons it is important that each person handles *only* his or her own blood, especially in the stages prior to fixation, or else handles blood from a known safe source. For the same reason, care is needed to minimize or eliminate spillage. Any spills should be cleaned up immediately with hypochlorite by the person concerned; and used culture tubes, syringes, pipettes etc. should be discarded into thick, plastic bags for proper disposal. Hygiene should be monitored regularly to ensure that the very small added risk is indeed kept very small.

Syringes

Special care is needed in the handling of syringes, to avoid inflicting injury and to avoid their subsequent misuse. Account for every syringe at issue and again after use.

Culture medium

Blood cells are cultured in a defined sterile medium containing amino acids, salts and vitamins, to which is added serum, antibiotics, phytohaemagglutinin and usually phenol red to act as a colour indicator. Making up an appropriate culture medium is difficult and time consuming and is thus not recommended. Culture medium complete with additives, and indeed entire culture kits, can be purchased from Biological Supply Houses at reasonable cost and these are excellent for class use. The medium is purchased as a dried powder and reconstituted with distilled water and then sterilized. Full directions for doing this and for adding serum, antibiotics (optional) and phytohaemagglutinin are provided with the dried powder. Alternatively, arrangements may be made for a suitable culture medium to be provided by the cytogenetics unit of a nearby hospital routinely involved in human chromosome analysis (with appropriate reimbursement for costs). This is the ideal approach.

The following items of equipment and materials will be required:

Sterile 10 ml plastic disposable syringe with no. 20 or 21 gauge needle or there-abouts; one per student. 75% ethyl alcohol and swab.

20 ml or thereabouts sterile, plastic, screw capped universal container with 9.5 ml of sterile culture medium and additives; one per student.

Incubator or water bath set at 37°C.

Centrifuge capable of 1000 r.p.m. or thereabouts; graduated 10 ml centrifuge tubes.

Sterile pasteur pipettes with bulbs; two per student. Ideally these should be *siliconized* by sucking up and then blowing out a little liquid silicone and then allowing the pipette to dry. The silicone helps to prevent cells sticking to the glass of the pipette.

Colchicine solution (0.001% = 10 µg/ml in distilled water); stored in a refrigerator.

Hypotonic (0.075 M KCl made up in distilled water), pre-warmed to 37°C.

Absolute methanol (preferably analytical grade) and glacial acetic acid, pre-cooled in a refrigerator.

Clean microscope slides; 95% alcohol.

Phosphate buffer, pH 6.8. This is made by mixing 30 ml each of 0.5 M potassium dihydrogen phosphate (KH_2PO_4) and 0.5 M disodium hydrogen phosphate (Na_2HPO_4), checking the pH and adjusting it if necessary with NaOH or HCl. Alternatively, buffer tablets can be purchased from Biological Suppy Houses and made up according to the directions provided.

Giemsa staining solution, prepared by adding 20 ml of stock to 100 ml distilled water to produce a 20% working solution.

The method is as follows:

1 *Collection of blood*. Arrange for a qualified medical person to collect 5–10 ml of blood separately from each student concerned. If making chromosome preparations from individual students is not desirable because of health or other reasons, obtain blood from a known male and female source. The blood should be collected aseptically into heparinized vacutainers and used immediately or stored in the refrigerator (4°C) for use on the following day.

2 *Setting up cultures*. Label a culture container on the *cap* (for ease of identification in the centrifuge). Loosen the cap. Swab the top of the vacutainer containing blood with 75% alcohol. Prepare a syringe.

   Mix the blood in the vacutainer by repeated inversion. Draw 2 ml of sterile air into the syringe by holding the needle tip close to (but not *in*) a bunsen flame. Then force the syringe needle through the rubber cap of the vacutainer and inject the air. Invert the tube and draw 2 ml of blood into the syringe. Remove the syringe needle from the vacutainer.

Replace the cap of the syringe and disconnect the needle plus cap, being careful not to touch the connection (to avoid contamination). Dispose of the needle and cap safely. Remove the top of the culture container and inject 0.5 ml (6–8 drops) of blood; recap the culture vessel. Dispose of the syringe and vacutainer safely. Shake the culture vessel gently and incubate it at 37°C for 72 h.

In the meantime rinse five slides in fresh 95% ethyl alcohol and then dry them with clean linen or tissue paper. Package the slides in clean typing paper or equivalent and place them in a freezer.

3 *Harvesting cultures.* After 72 h incubation add to the culture 0.1 ml of colchicine solution, mix well and incubate for a further 1 h. Omit the colchicine if you wish to study *mitosis* in human cells.

Label a centrifuge tube (using a code is easiest). Shake the culture well and transfer its entire contents to the centrifuge tube. Balance the centrifuge and run at 1000 r.p.m. or thereabouts for 10 min.

Label a sterile pipette. Remove the centrifuge tube carefully and gently pipette off and discard the supernatant to 1 ml. Note that the cells we are seeking are in the thin opaque layer resting on the red blood cells that form the bulk of the 'cell button' at the base of the centrifuge tube. In this and subsequent steps *be careful not to disrupt the cell button* until directed.

Resuspend the cells in the remaining liquid by blowing in air with a pipette. Gradually add 5 ml pre-warmed hypotonic solution (0.075 M KCl), mix well with the pipette and incubate for 20 min at 37°C.

In the meantime prepare cold fixative (it must be fresh and cold) by mixing 3 parts absolute methanol with 1 part glacial acetic acid. Hold the fixative at all times between use in the refrigerator.

After 20 min incubation in the hypotonic solution, centrifuge as before. Draw off the supernatant and resuspend the cells in the dregs of KCl.

Add 5 ml cold fixative to the culture by squirting in with your pipette, mixing well. Centrifuge as before.

Discard the supernatant to 1 ml, resuspend cells in the remaining liquid and then add a second 5 ml lot of fixative, mixing well. Leave overnight if necessary. Centrifuge as before.

Repeat the previous step. Discard the supernatant to 0.5–1 ml, depending on the size of the button (which by now will be almost colourless and difficult to see; do not be discouraged!). Resuspend cells.

4 *Making slides.* (First practise the following with unwanted fixative and spare slides.) Take up some of the suspension into a fresh pipette. Hold the pipette about 25 cm above a frosted slide with the slide sloping downwards at an angle of about 65°. This is where you actually apply the chromosomes: do so by dropping 3 drops of suspension on to separate areas of the slide, allowing each drop to run down the slide before adding another. Ignite the alcohol on the slide with a bunsen flame and allow it to burn off.

Allow slides to dry overnight (longer if necessary).

5 *Staining.* Immerse slide in 20% Giemsa for 5 min. Wash off excess stain with phosphate buffer. Gently blot dry with filter paper.

Leave slides overnight in a dust-free environment to dry completely. Mount slides in DPX with long, thin (No.1) coverslips.

EXERCISES

1 Common Human Traits

This exercise simply involves recording the phenotype of each student in a class with respect to PTC tasting, hair whorl rotation and tongue rolling ability; and/ or other traits such as those listed in Table 9.1.

For PTC tasting, a class of 30 can easily be tested over a half-hour period, during which students can also be deciding other aspects of their phenotype. It is best to advise students to wash out their mouth after the PTC test before swallowing. PTC tasting may not be permitted in some areas, because of possible health concerns. The bitter taste of PTC may be disconcerting to some.

For tongue rolling, it is best to allow a few days' practice before a student finally decides he or she is a non-roller.

For hair whorl rotation, it is easiest to have a colleague record one's phenotype; or to use a mirror! The trait is more difficult to score in females because of hairstyle; combing the hair away from the crown helps.

After phenotyping each student, pool the data obtained and calculate phenotype frequencies, males and females and/or racial groups separately for added interest. Comparisons can be made with data collected from previous classes or with published data such as those available for PTC tasting.

Actual class data are shown in Table 9.3.

Table 9.3 Numbers and proportions of different phenotypes of four human traits found in 52 females and 67 males from a first-year university genetics class.

Trait	Phenotype	Female		Male	
		No.	Proportion	No.	Proportion
PTC tasting	Taster	40	0.77	43	0.64
	Non-taster	12	0.23	24	0.36
Hair whorl rotation	Clockwise	43	0.83	48	0.72
	Counter-clockwise	9	0.17	19	0.28
Tongue rolling ability	Roller	37	0.71	52	0.78
	Non-roller	15	0.29	15	0.22

Genetic variation in humans

One purpose of this exercise is to introduce students to the range of human traits that have a genetic component, involving biochemical traits such as tasting, morphological traits such as hair whorl rotation, and behavioural traits such as tongue rolling. Individually, students become aware of and thus greatly interested in their phenotype (and hence their possible genotype), especially if a wide range of traits (Table 9.1) are examined. Collectively, students become motivated by the extent of genetic variability present in human populations and the manner in which genetic traits vary between racial groups and with sex. Any suspected differences can be tested for using the contingency chi-squared method (e.g. Table 9.4). The study can be concluded with a discussion on the possible roles of learning, cultural and other environmental factors in the development of the traits studied; and on possible adaptive significance of racial differences such as those for PTC tasting (Table 9.5 and following section).

Table 9.4 Comparison by numbers and proportions of female and male PTC tasters and non-tasters; and results of a contingency chi-squared test. Data from Table 9.3.

	Tasters	Non-tasters	Total
Females	40 (0.77)	12 (0.23)	52
Males	43 (0.64)	24 (0.36)	67

Contingency $\chi^2 = 2.2$, $d.f.= 1$, $P = 0.15$ (not significant).

Table 9.5 Proportion of PTC tasters in various human populations. Data mostly from Lerner (1968).

Population	Proportion of tasters
Australian Aborigines	0.27
Eskimos	0.59
Caucasians	0.69
New Zealand Maori	0.90
Chinese	0.93
North American Indians	0.97

Gene and genotype frequency estimates

Data such as those of Table 9.3 can be used as models to estimate gene and genotype frequencies in the population concerned, using the Hardy-Weinberg method. In doing this we assume that the traits are each determined by a single gene with a pair of alleles, one of which is completely dominant (see p.172). Using the class data we proceed as follows.

Let p = the frequency of the dominant allele (say **T** for PTC tasting) and
let q = the frequency of the recessive allele (**t**). Then
$p + q$ = 1   (eqn 1)

If the population is in Hardy-Weinberg equilibrium then its genotypic composition is given by the binomial expansion of $(p + q)^2$, which is
$p^2 + 2pq + q^2$, where p^2 = the frequency of the homozygous dominant genotype (say **TT**)
$2pq$ = the frequency of the heterozygote (**Tt**) and
q^2 = the frequency of the homozygous recessive (**tt**).

The square root of the frequency of the double recessive condition, therefore, gives us an estimate of the frequency of the recessive allele in the population, from which an estimate of the frequency of the dominant allele is obtained by subtraction. Thus for the PTC data of Table 9.3,

q (female) = $\sqrt{0.23} = 0.48$. Therefore $p = 1 - 0.48 = 0.52$ (from eqn 1 above).
q (male)   = $\sqrt{0.36} = 0.60$. Therefore $p = 1 - 0.60 = 0.40$.

The data are summarized in Table 9.6, where estimates of the proportion of homozygous dominant and heterozygous individuals in the population are also given.

Table 9.6 Gene and genotype frequency estimates for PTC tasting, calculated from the data of Table 9.3 and assuming the population was in Hardy-Weinberg equilibrium.

| | Gene (allele) | | Genotype | | |
	T $(=p)$	t $(=q)$	TT $(=p^2)$	Tt $(=2pq)$	tt $(=q^2)$
Female	0.52	0.48	0.27	0.50	0.23
Male	0.40	0.60	0.16	0.48	0.36

We have, of course, assumed in the above calculations that the population we sampled was in Hardy-Weinberg equilibrium, implying that tasting or non-tasting has no effect upon survival (viability), reproductive capacity or choice of mate (i.e. there is no selection for or against either phenotype and mating between the two phenotypes is random). At least in the past history of humans this may not have been the case, for otherwise it is difficult to explain the phenotype frequency differences for PTC tasting between different racial groups (Table 9.5). Non-tasters are also more frequent than expected amongst individuals afflicted with certain thyroid disorders. Perhaps PTC tasting somehow protects against thyroid disease.

Similar calculations and comparisons can be made for the other two traits examined in Table 9.3, or any other trait for which monohybrid inheritance with dominance is assumed.

2 ABO Blood Groups

This exercise involves obtaining the ABO blood group type of members of a class, either by testing in the laboratory or, if this is not permitted, by having the test done by a qualified medical person. Pool the data obtained and express blood group frequencies as percentages of the total.

Actual results from a class of 137 students are shown in Table 9.7.

Table 9.7 Numbers and proportions of ABO phenotypes in a class of 137 students from Wellington, NZ.

| | Phenotype (blood group) | | | | |
	A	B	AB	O	Total
Number	55	19	4	59	137
Proportion	0.40	0.14	0.03	0.43	1.00

Data such as these may be used for a number of different purposes in a practical genetics class, depending on the level of teaching.

Understanding the genetics of the ABO blood groups

This exercise leads us firstly to an understanding of the origin of the four ABO phenotypes, deriving as they do from the existence of *three* allelic forms of the I gene. Thus we are introduced to the phenomenon of *multiple* allelism – i.e. the existence in some (perhaps all) cases of not two or a few but of many forms of a particular gene. Moreover, the existence of the **AB** phenotype can be understood only if I^A and I^B alleles are co-dominant, rather than one being dominant and the other recessive; while the existence of four rather than six phenotypes is understood only if the i allele is recessive to both I^A and I^B. These different dominance relationships amongst the three alleles can be easily understood in terms of production or non-production of their gene products. Thus the I^A and I^B alleles specify the formation of A and B antigens, respectively, while the recessive i allele produces no antigen.

Phenotype frequencies in different populations

The data of Table 9.7 show that the four blood group phenotypes do not occur in equal frequencies – phenotypes **A** and **O** are much more common than the **B** phenotype. This inequality has been found in all human populations tested so far. Moreover, racial groups differ in their relative proportions of the four phenotypes, as shown in Table 9.8.

Table 9.8 Proportions of ABO blood group phenotypes in selected populations. *Sources:* abridged from Stern (1973), Woodfield *et al.* (1987) and Sutton (1988).

Population	*A*	*Phenotype (blood group)* *B*	*AB*	*O*
Chinese	0.27	0.23	0.06	0.44
French	0.45	0.09	0.04	0.42
Indians	0.25	0.38	0.07	0.30
Nigerians	0.21	0.23	0.04	0.52
New Zealand European	0.40	0.10	0.03	0.47
New Zealand Maori	0.52	0.04	0.01	0.43
Amerindians (Navaho)	0.22	0.00	0.00	0.78
US 'blacks'	0.27	0.21	0.04	0.48
US 'whites'	0.41	0.10	0.04	0.45
Welsh	0.35	0.10	0.03	0.52

Results obtained in a class exercise should be examined to see whether or not they are representative of the local or other population(s), using either a standard or a contingency chi-squared test. Table 9.9 compares the class data of Table 9.7 with those representative of N.Z. Europeans (Table 9.8).

From Table 9.9 we conclude that the class data of Table 9.7 are indeed representative of the population as a whole from which they were obtained. If a significant difference had been found we would need to consider by way of explanation such things as sample size or composition of the class (in which an atypical mix of Asians and Europeans, for example, may lead to unexpected results).

Table 9.9 ABO blood group comparison.

	A	B	AB	O	Total
Observed (from Table 9.7)	55	19	4	59	137
Expected (calculated from the proportions given in Table 9.8)	54.8	13.7	4.1	64.4	137

$\chi^2 = 2.5$, $d.f. = 3$, $P = 0.5$. Conclusion: there are no significant differences between the observed and expected frequencies of the ABO phenotypes.

Gene and genotype frequency estimates

Estimates of gene frequencies can be obtained from the data of Tables 9.7 and 9.8. Since we are dealing with *three* alleles (rather than the usual two alleles, as in Exercise 1), we use the *trinomial* expansion of $(p + q + r)^2$, where p = the frequency of I^A, q = the frequency of I^B, and r = the frequency of i; and where

$$p + q + r = 1 \qquad\qquad (\text{eqn 2})$$

The expansion is $p^2 + 2pr + q^2 + 2qr + 2pq + r^2$. The frequencies of the four ABO phenotypes and genotypes are as shown in Table 9.10.

Table 9.10 Blood group frequencies as given by the expansion of $(p + q + r)^2$.

	Phenotype			
	A	B	AB	O
Genotype	$I^A I^A$ plus $I^A i$	$I^B I^B$ plus $I^B i$	$I^A I^B$	$i i$
Frequency	$p^2 + 2pr$	$q^2 + 2qr$	$2pq$	r^2

Estimates of the frequencies of the three alleles, using the data of Table 9.7, are obtained as follows:

1 Frequency of i (= r) = *the square root of the frequency of blood group O*
                         = $\sqrt{r^2}$
                         = $\sqrt{0.43}$
                         = 0.66.

2 Frequency of I^A (=p) = *1 minus the square root of the frequencies of blood groups B and O combined* (see p. 186)*
                            = $1 - \sqrt{(0.14 + 0.43)}$
                            = $1 - \sqrt{0.57}$
                            = 0.25.

3 Frequency of I^B (=q) = *1 minus the square root of the frequencies of blood groups A and O combined* (see p. 186)*
                            = $1 - \sqrt{(0.40 + 0.43)}$
                            = $1 - \sqrt{0.83}$
                            = 0.09.

$$Check: (p + q + r) \quad = \quad 0.25 + 0.09 + 0.66$$
$$= \quad 1.0^{+}$$

These gene frequencies can be used to estimate the numbers of homozygotes and heterozygotes expected in our class of 137 (= *N*), as shown in Table 9.11.

Table 9.11 Expected number of ABO genotypes in a class of 137 students, based on gene frequency calculations.

Genotype	Frequency	Formula	Calculation	Expected number in class
$I^A I^A$	p^2	$p^2 \times N$	0.25 x 0.25 x 137	9
$I^A i$	$2pr$	$2pr \times N$	2 x 0.25 x 0.66 x 137	45
$I^B I^B$	q^2	$q^2 \times N$	0.09 x 0.09 x 137	1
$I^B i$	$2qr$	$2qr \times N$	2 x 0.09 x 0.66 x 137	16
$I^A I^B$	$2pq$	$2pq \times N$	2 x 0.25 x 0.09 x 137	6
ii	r^2	$r^2 \times N$	0.66 x 0.66 x 137	60
Total				137

These estimates are, of course, based on Hardy-Weinberg assumptions, including random mating amongst members having different blood groups. They give an expected number of heterozygotes $I^A I^B$ (blood group **AB**) of 6 in our class of 137, which is reasonably close to the observed number of 3. This suggests that our population does indeed conform to Hardy-Weinberg assumptions implying, amongst other things, little or no selection (today at least!) for any one phenotype, and random breeding amongst members carrying different phenotypes.

3 Human chromosomes

This exercise involves the preparation and study of the human karyotype. Students may work from their own slides, or from ready-made demonstration

* This is because

$$p \quad = \quad 1 - q - r \text{ (from eqn 2)}$$
$$= \quad 1 - (q + r). \text{ Now, we can solve for } (q + r) \text{ because}$$
$$(q + r) \quad = \quad \sqrt{(q + r)^2}$$
$$= \quad \sqrt{(q^2 + 2qr + r^2)} \text{ by expansion. Substituting we get}$$

Therefore

$$p \quad = \quad 1 - \sqrt{(q^2 + 2qr + r^2)}.$$

Similarly,

$$q \quad = \quad 1 - p - r$$
$$= \quad 1 - (p + r).$$
$$(p + r) \quad = \quad \sqrt{(p + r)^2}$$
$$= \quad \sqrt{(p^2 + 2pr + r^2)}$$

Therefore

$$q \quad = \quad 1 - \sqrt{(p^2 + 2pr + r^2)}.$$

+ The estimated values of *p, q* and *r* sometimes do not add to exactly 1.0. For present purposes any discrepancy can be removed by dividing the difference between observed and expected proportionately between the three gene frequencies.

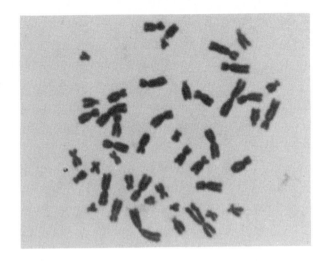

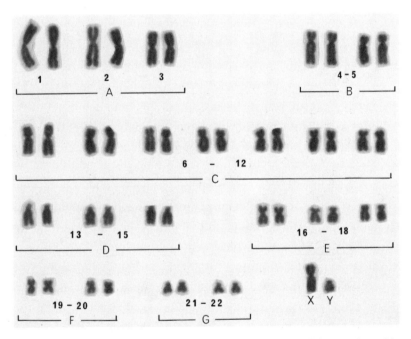

Fig. 9.3 Above: chromosomes of a human male prepared from cultured lymphocytes by the technique described in this chapter. Each chromosome is in fact a double structure (i.e. composed of two chromatids) because the cell was undergoing mitosis at the time it was fixed. The centromere is located at the constriction (waist) of the chromosome, and varies in position from median (metacentric) to subterminal (acrocentric).

Below: the same chromosomes as above but after they were cut out from a photograph, arranged in pairs according to size and centromere position and pasted onto a card, so as to form a karyotype. The chromosome pairs are numbered 1 – 22 plus XY and arranged into seven conventional groups. Within each group the pairing and numbering is arbitrary because specific chromosomes cannot be distinguished with certainty, except in group A where differences in length and centromere position normally enable chromosomes 1 – 3 to be separately identified.

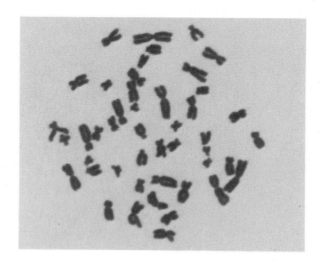

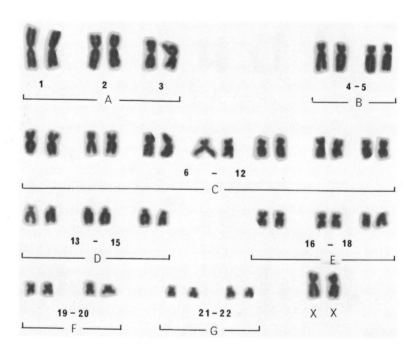

Fig. 9.4 As for Fig. 9.3 except from a female, in which there are two X chromosomes.

slides, or from photographs. Slides made in class should be scanned systematically for good chromosome spreads (Figs 9.3 and 9.4). Often certain regions of a slide will have better spreads than others, due to differential heating of the slide during flaming. Suitable spreads from these slides, or from demonstration slides, should be drawn with a camera lucida, or photographed. After photography, prints can be cut out and glued to prepared cards for the construction of a karyotype (Figs 9.3 and 9.4). Observations of *mitosis* in human cells can be made from slides prepared from cultures to which colchicine was not added (Fig. 9.5).

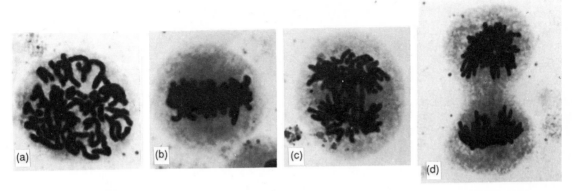

Fig. 9.5 Mitosis in cultured human lymphocytes prepared according to the technique described in this chapter except that treatment with colchicine prior to harvesting of cultures was omitted. (a) prophase, (b) metaphase, (c) anaphase, (d) telophase. At telophase the beginnings of the cleavage furrow can be seen between the two groups of chromosomes.

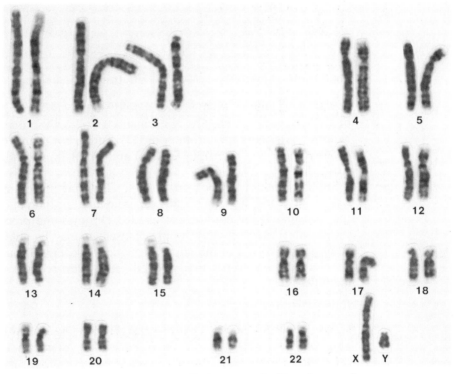

Fig. 9.6 G-banded chromosomes of a human male. Homologous chromosomes are paired and numbered 1 - 22 plus XY. The banding patterns allow separate identification of each chromosome pair. Preparation and photograph per courtesy of D. Romain, Cytogenetics Unit, Wellington Hospital (NZ).

If working from ready-made demonstration slides or photographs (which can be purchased as kits from some Biological Supply Houses), karyotypes of a

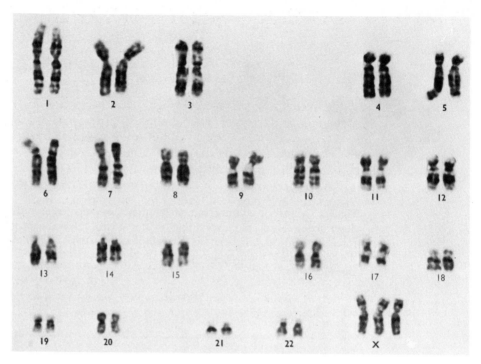

Fig. 9.7 As for Fig. 9.6 except from an XXX (triple X) individual.

normal male and a normal female should be prepared. In this way also, a number of well-known chromosome abnormalities such as those associated with Down syndrome (trisomy for chromosome 21), Turner syndrome (XO), Klinefelter syndrome (XXY); and the triple X (XXX) and 14/21 translocation conditions can be studied.

Students should also be introduced to *G-banding* procedures, which are used in diagnostic work and which allow separate identification of every chromosome of the human set. Photographs of banded chromosome preparations can also be purchased from Biological Supply Houses, or may be obtained from a hospital laboratory involved in human cytogenetic work (Figs 9.6 and 9.7). These should be studied to confirm that, indeed, each chromosome can be individually identified.

REFERENCES

Lerner I.M. (1968). *Heredity, Evolution and Society.* San Francisco, W.H. Freeman & Co.

McKusick V.A. (1986). *Mendelian Inheritance in Man.* 7th edition. Philadelphia, The Johns Hopkins University Press.

Priest J.H. (1977). *Medical Cytogenetics and Cell Culture.* 2nd edition. Baltimore, Lea & Febiger.

Stern C. (1973). *Principles of Human Genetics.* 3rd edition. San Francisco, W.H. Freeman & Co.

Sutton H.E. (1980). *An Introduction to Human Genetics.* 4th edition. New York, Harcourt Brace Jovanovich Inc.

Woodfield D.G., Simpson L.A., Seber G.A.F. and McInerney P.J. (1987). Blood groups and other genetic markers in New Zealand European and Maoris. *Annals of Human Biology,* **14,** 29–37.

Chapter 10 GENETICS WITH BACTERIA

Experiments with micro-organisms such as bacteria are technically more demanding, in terms of facilities and equipment required, than are those with eukaryotic organisms. Work with micro-organisms also requires much stricter control over safety procedures, and involves a higher level of expenditure on consumable items. Information on safety aspects of handling micro-organisms in a classroom situation is included in the text to this chapter as well as in the Appendix section on Safe Handling of Micro-organisms. The main safety requirements are to use sterile techniques at all times, to avoid contamination of culture media and culture vessels by potentially pathogenic organisms, and to dispose of all used materials safely by autoclaving after completion of an exercise.

The exercises given in this chapter are more appropriate to undergraduate university or polytechnic classes than to schools or colleges; and at the level for which this book is written exercises involving bacterial viruses were considered to be not appropriate.

INTRODUCTION AND BACKGROUND

Studies with bacteria have made a huge contribution to the science of genetics. Once the technical difficulties of working with these tiny organisms were overcome in the 1950s they rapidly proved to be very useful in tackling a whole range of problems concerning gene structure and organization, gene mutation, gene regulation and action, and many aspects of what later developed as the study of molecular genetics. The value of micro-organisms as experimental systems continues to develop and now they are key organisms in the study and practical application of recombinant DNA technology and genetic engineering. Some basic experience in handling bacteria is therefore clearly desirable in practical genetics classes.

Bacteria are *prokaryotes*. Thus they differ from all of the other organisms (i.e. eukaryotes) that we have dealt with so far in this book. Prokaryotes lack a true nucleus, in that their genetic material is not separated from the cytoplasm by a membrane, and their chromosomes are simply naked molecules of double-stranded DNA. They also lack cell organelles such as chloroplasts and mitochondria. Bacteria have a single chromosome which replicates in a semi-conservative way and is distributed to the daughter cells during cell division. There is no mitosis or meiosis as we find in eukaryotes. The genetic system of bacteria, therefore, is somewhat different from that which we find in other organisms, and this affects the way we study and experiment with these organisms.

Apart from their developmental and biochemical simplicity the other advantages of bacteria include their short life cycle, which can be as little as 40 min, and the facts that we can handle them by the million and use selective methods to detect mutants and rare recombinants. Bacteria are also essentially haploid, and recessive mutations therefore show up readily in the phenotype.

Mating and Exchange of Genetic Material in Bacteria

In common with most other organisms, bacteria mate and exchange and recombine their genetic material. There are three such processes of exchange, viz. *transformation, transduction* and *conjugation.* Transformation is the exchange of genetic information brought about by the simple uptake of naked DNA by a recipient cell. Quite large pieces of DNA can pass through the bacterial cell wall and then undergo recombination with homologous parts of the recipient's chromosome. In transduction the transfer of genetic material from one bacterium to another is brought about by a virus.

The main exercise described in this chapter concerns the third process, conjugation, for which we will therefore present more detail. The experimental organism we will be concerned with is the bacterium *Escherichia coli (E. coli),* which is a normal inhabitant of the intestine of humans.

Conjugation in bacteria is defined as the one-way transfer of genetic information from a donor cell to a recipient cell involving direct contact of cells. In *E. coli* the transfer takes place between cells of opposite mating types. Mating type in *E. coli* is determined by a small, circular piece of DNA called a plasmid, which is known as the *sex factor* or *fertility factor,* and is symbolized **F.** The F-factor is about $1/40$ the size of the bacterium's chromosome. Cells carrying this factor are donors and are designated **F+**, while those lacking the F-factor are recipients and known as **F-**. In donor cells the F-factor can exist in one of two alternative states: it may be free in the cytoplasm and behave independently of the main chromosome, or it may be integrated into the chromosome and replicated and transmitted along with it. Strains which carry the F-factor in this integrated form are known as **Hfr**

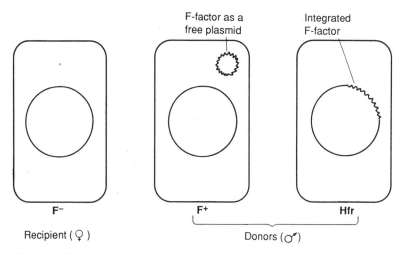

Fig. 10.1 Illustrations of mating types in *E. coli ,* as determined by the sex factor, **F.** Cells lacking the F-factor are designated as F- (females), and they receive genetic material from F+ and Hfr types which are the donors (males). The bacterial chromosome and the F-factor are shown as single lines for simplicity.

strains (**Hfr** = high frequency of recombination, so named because these strains mate and exchange genetic material at very high frequencies). The relationship between the three cell types with respect to the sex factor is shown in Fig. 10.1. The **F⁺** and **Hfr** states are reversible, the change from one to the other being brought about by a recombination event between the F-factor and a homologous region of the main chromosome, which excises or integrates the sex plasmid. The F-factor may integrate at various locations within the main chromosome, which gives rise to a number of different **Hfr** strains, such as the **HfrH** strain of Exercise 2.

During conjugation there is mating between **F⁺** and **F⁻** strains, which involves the formation of a cytoplasmic bridge connecting the pair of cells involved. The way in which genetic material is exchanged depends upon the type of donor cell, i.e. whether the donor is **F⁺** or **Hfr**.

The F⁺ x F⁻ cross

In this cross the F-factor in the F⁺ cell replicates and as it does so a copy of it is transferred to the recipient F⁻ cell, which then becomes F⁺. During this event there is no exchange of material with the main bacterial chromosome (Fig. 10.2).

The Hfr x F⁻ cross

This cross is the one which is most useful in experimentation because it involves transfer of genes on the chromosome from the donor to the recipient cell, and can be used in recombination analysis. The process is illustrated and described in Fig. 10.3. Because of the linear way in which chromosome transfer takes place, starting at one end, it is possible to use this system to order genes along the chromosome. This is done by an interrupted mating procedure in which conjugating pairs of cells are sampled at intervals of time and then separated artificially by violent shaking. The order in the time sequence in which marker genes from the donor appear in the progeny of the cross gives their linear order within the chromosome.

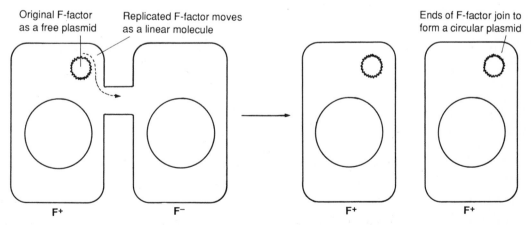

Fig. 10.2 Diagram showing conjugation between **F⁺** and **F⁻** cells of *E. coli*. In this mating process the F-factor in the male donor cell replicates, and a copy of it is then donated to the F⁻ recipient, which then also becomes an **F⁺** cell.

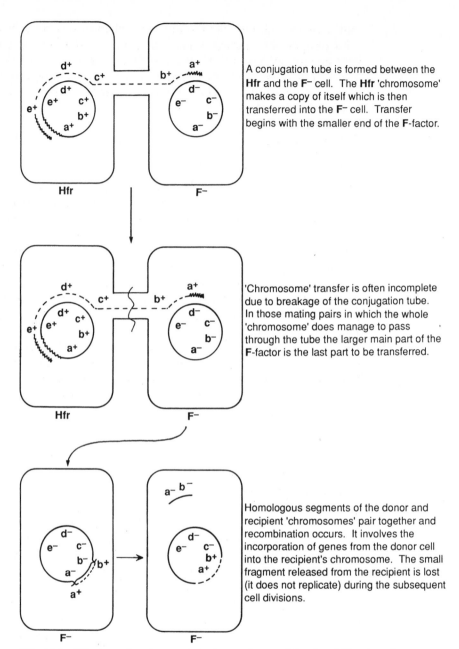

A conjugation tube is formed between the **Hfr** and the **F⁻** cell. The **Hfr** 'chromosome' makes a copy of itself which is then transferred into the **F⁻** cell. Transfer begins with the smaller end of the **F**-factor.

'Chromosome' transfer is often incomplete due to breakage of the conjugation tube. In those mating pairs in which the whole 'chromosome' does manage to pass through the tube the larger main part of the **F**-factor is the last part to be transferred.

Homologous segments of the donor and recipient 'chromosomes' pair together and recombination occurs. It involves the incorporation of genes from the donor cell into the recipient's chromosome. The small fragment released from the recipient is lost (it does not replicate) during the subsequent cell divisions.

Fig. 10.3 Diagram showing conjugation and recombination following chromosome transfer between **Hfr** and **F⁻** cells of *E. coli*. Chromosomes are shown as single strands for simplicity, and the symbols **a, b, c** etc. represent marker genes in either their wild type (+) or mutant (-) forms.

Gene symbols

Standard symbols are used to describe various mutants and their normal (i.e. wild type) strains. Genes are named by three-letter symbols which are written in italics and which indicate the biochemical functions controlled by the genes. A minus superscript indicates loss of activity of the gene concerned, and a plus superscript indicates a wild type activity: e.g. *leu⁻*, *pro⁻* and *thi⁻* indicate inability to synthesize leucine, proline and thiamin, respectively; and *leu⁺*, *pro⁺* and *thi⁺* are the wild type alleles which have the capacity to make these substances.

PROCEDURES

Use of Sterile Technique

The use of sterile technique is critical in all work with bacteria. The growth media used are highly nutritious and there is the ever present hazard of picking up unwanted and potentially dangerous organisms from outside sources, as well as the nuisance of spoiling work by contamination. For convenience as well as safety it is essential to follow the guidelines on the safe handling of micro-organisms which are given in the Appendix, and to pay particular attention to sterilizing all work surfaces and items of equipment. Media and all other sterile materials should be uncovered for the minimum of time necessary for working. Particular care should be taken both with pipettes, to ensure that the tip is kept out of contact with non-sterile surfaces, and with sterile screw-capped bottles during the opening and transfer stages. It is standard procedure to flame the caps of bottles before removal, and then to flame the mouth before and after use. Wire loops and straight wire needles are sterilized by heating to red-hot with a bunsen flame, and then cooling for a few seconds before use.

Keeping Stock Cultures of Bacteria

Bacterial strains are kept as stock cultures in small 2 ml glass tubes containing 'stab' medium. The tubes are sealed with plastic stoppers. Prepared tubes are inoculated by picking up cells from a colony of the culture concerned with a sterile straight wire and stabbing the wire into the medium in the tube. The tubes are then incubated at 37°C overnight before being stored in the dark at room temperature. Stock cultures prepared in this way will keep for 6 months, after which they should be revived on plates of nutrient agar, re-isolated from single colonies and then returned to fresh storage tubes.

Purifying a Bacterial Culture

In order to purify bacterial strains, as in the preparation of stocks to go into storage, it is sometimes necesary to grow strains out on plates of nutrient agar in such a way that a sub-culture can be taken from a single colony. Single colonies originate from single cells, and when inoculum is taken from such a colony it produces a new growth of cells which are genetically pure and which have a high probability of belonging to the strain from which they originated. A sterile wire loop is used to pick up a small amount of inoculum from a culture, or a colony, and is streaked several times across one side of the agar plate, as shown in Fig. 10.4. The loop is then flamed off to remove surplus cells, cooled and then streaked several times at right angles across the first set of streaks. The idea is to

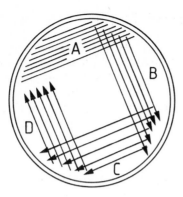

Fig. 10.4 Diagram showing how to streak out a culture plate for the isolation of bacteria derived from a single colony.

thin out the inoculum progressively, as shown in Fig. 10.4, until the final streaks are dilute enough to spread out as single cells, as in area D. Following incubation at 37°C for 18 h (with the plates inverted to avoid condensation accumulating) several clearly separated colonies will be found in area D. One of these with the appearance typical of the strain can be used to start a new pure culture.

Overnight Cultures of Bacteria

Experimental protocols usually start with an overnight bacterial culture. To prepare such a culture a plate is first produced from the required stock culture and an inoculum from a single colony is then taken into 5 ml of broth in a 15 ml screw-capped bottle. The culture is then incubated with agitation for 18 h at 37°C (i.e. overnight). It will grow to give a concentration of about 10^9 cells per ml.

Exponential Cultures of Bacteria

Exponentially growing cultures are made in order to provide students with high cell densities at the start of an experiment. Such a culture is produced by 1 : 50 dilution of an overnight culture into fresh medium (broth) in a small screw-capped bottle. The culture is then incubated for $1\frac{1}{2}$ h on a rotary shaker running at 33 r.p.m. to produce an exponential culture with about 2×10^8 cells/ml.

Dilutions

Cultures of bacteria are used at high concentrations, usually with as many as 10^8 or 10^9 cells per ml. To estimate the number of colony-forming cells in a given culture it is necessary for the culture to be diluted to manageable concentrations. For this purpose serial dilutions are made in the way already described for asco-spore dilutions in Aspergillus in Chapter 6. Tenfold or hundredfold dilutions may be made using buffer, and a fresh sterile pipette is required at each stage of dilution.

Assaying Numbers of Bacterial Cells

To assay numbers of viable, colony-forming cells a liquid culture of bacteria is first diluted to a concentration of between 1000 and 5000 cells/ml. A 0.1 ml drop (i.e. 100–500 cells) of this dilution is then placed via a pipette on an agar plate of nutrient medium and spread with a glass spreader. The spreader is simply a bent glass rod which is sterilized by immersion in alcohol and then flaming. Plates are incubated for 18 h at 37°C and colony numbers then counted.

Preparing Buffers, Media and Culture Plates

Buffer

Na_2HPO_4 (anhydrous)	7 g
KH_2PO_4	3 g
NaCl	4 g
$MgSO_4.7H_2O$	0.2 g
Water	1 litre

Dissolve the salts in the order given; dispense into 100 ml screw-capped bottles and sterilize at 15 p.s.i. for 15 min.

Stab medium

Difco nutrient broth powder	0.9 g
NaCl	0.5 g
Difco 'Bacto' agar powder	0.75 g
Water	100 ml

Add the agar to cold water and dissolve by heating for 30 min; then add and dissolve the other substances. Sterilize at 15 p.s.i. for 15 min and then dispense into sterile tubes to half-full. Cap the tubes.

Nutrient broth (= 'broth')

Oxoid No. 2 nutrient broth powder	25 g
Water	1 litre

Dissolve the powder and dispense into screw-capped bottles of a convenient size. Sterilize at 15 p.s.i. for 15 min.

Nutrient agar

Oxoid No. 2 nutrient broth powder	25 g
Davis New Zealand agar	12.5 g
Water	1 litre

Dissolve ingredients by heating in water; dispense into 500 ml bottles and sterilize at 15 p.s.i. for 20 min.

Minimal agar

Water agar	300 ml
Minimal salts (x4 concentrate)	100 ml
20% glucose	4 ml

Melt stock water agar (see Chapter 6) at 100°C for 15 min. Then add warmed sterile salts and glucose, mix and pour the plates.

Minimal salts (x4 concentrate)

NH_4Cl	20 g
NH_4NO_3	4 g
Na_2SO_4 (anhydrous)	8 g
K_2HPO_4 (anhydrous)	12 g
KH_2PO_4	4 g
$MgSO_4.7H_2O$	0.4 g
Water	1 litre

Dissolve salts in the order given. Filter into 100 ml bottles and sterilize at 15 p.s.i. for 15 min.

20% glucose (x 100 concentrate)

D-glucose	200 g
Water	1 litre

Dissolve glucose; dispense into 100 ml bottles and sterilize at 15 p.s.i. for 15 min.

Soft agar

Difco 'Bacto' agar	6 g
Water	1 litre

Dissolve agar by heating in water and dispense into bottles (50, 100ml). Sterilize at 15 p.s.i. for 15 min.

Supplements, added to media to give concentrations of:

Amino acids	20 µg/ml
Vitamins	1 µg/ml
Streptomycin	200 µg/ml

Culture plates

Glass or sterile disposable petri dishes may be used. Glass dishes will need sterilizing and this can be done by wrapping them as packs of ten in paper or aluminium foil and then autoclaving at 15 p.s.i. for 20 min. Medium is poured while it is still warm (60–70°C), i.e. before it solidifies, or in the case of stored solid medium after melting at 121°C. Lids should be removed only at the time of pouring and about 25 ml are used for growing bacteria as separate colonies in petri dishes. Heating dishes over a bunsen flame will drive air bubbles out of the agar. Excessive condensation can be removed by drying the plates in an incubator overnight in an inverted position with their lids on.

EXERCISES

1 Growth Requirements of Wild Type and Mutants of *E. coli*

This simple exercise will introduce students to some of the techniques involved in handling *E. coli* and to the growth requirements (biochemical phenotypes) of wild type and some mutant strains. It can also serve as preparation for the more challenging Exercise 2, which involves mating and recombination. The procedures to use in this simple exercise are essentially the same as those described for yeast (Exercise 1 of Chapter 4) and involve inoculating wild type and selected mutant strains of *E. coli* onto complete medium (nutrient agar), minimal medium (minimal agar) and appropriate supplemented media. Strains suitable for this exercise are *E. coli* K12 (wild type) and one or more mutants such as *leu⁻*, *pro⁻* or *trp⁻*, alone or in combination. These mutants are unable to synthesize leucine, proline and tryptophan, respectively, and thus will not grow on minimal medium unless it is supplemented with the appropriate amino acid. For variation, a mutant requiring the vitamin thiamin (*thi⁻*), or streptomycin sensitive (wild type) and resistant (*st^R*) strains can be used. Appropriate strains can be obtained from a university or polytechnic department which routinely teaches bacterial genetics courses.

Requirements (per student)
Access to wild type and appropriate mutant strain(s) of *E. coli*, growing either in small bottles of nutrient broth or on nutrient agar. The cultures are best labelled as 'unknown 1, 2, etc.', so as not to disclose their genotype to students beforehand.
Nutrient agar plate.
Minimal agar plate.
Minimal agar supplemented with the compound (amino acid, vitamin, etc.) that the mutant(s) being used is unable to synthesize.
Wire loop, marker pen.

Method
1 Mark the underside of each agar plate into halves, thirds etc. (depending on the number of mutants being tested). Label the segments 'unknown 1', 'unknown 2', etc.
2 Using sterile technique throughout, transfer via a wire loop a small quantity of bacteria from each 'unknown' culture to its appropriate segment on each agar plate. Make sure the wire loop is properly re-sterilized between transfers; and spread out each inoculum over its area of agar.
3 Label the plates (name, exercise, etc.) and incubate them for 24 or 48 hours and observe. Classify the 'unknowns' according to which agar they grow on.
4 Interpret the results in biochemical–genetic terms as for Chapter 4, the only difference being that in *E. coli* there are no colour markers to distinguish between wild type and most biochemical mutants.

2 Recombination and Gene Mapping in *E. coli*

This exercise demonstrates the process of recombination in bacteria, and the order of linkage of genes in the chromosome. The exercise involves conjugation between **Hfr** and **F⁻** strains of *E. coli*. Exponentially growing cultures of the two strains are mixed together and incubated at 37°C. Mating takes place between pairs of cells, and at various intervals of time after the start of the exercise samples of the mating mixture are taken, diluted and shaken to break the conjugation tubes. The samples are then plated out onto differential media to select for different combinations of marker genes donated to the F⁻ cells by the **Hfr** males. The original **Hfr** strain is streptomycin sensitive (str^S) while the F⁻ strain is streptomycin resistant (str^R). Streptomycin is included in the supplemented media so that only recombinant F⁻ cells are recovered by sampling. Numbers of recombinants are counted for each sampling time and plotted on a graph to determine the time of entry of the marker genes, thus giving their order along the chromosome.

Strains suitable for this exercise are:

E. coli K12 **HfrH**   *leu⁺ pro⁺ trp⁺ str^S*
E. coli K12 **F⁻**   *leu⁻ pro⁻ trp⁻ str^R*

These strains can be obtained from a university or polytechnic department which routinely teaches bacterial genetics courses. There are a number of **Hfr** strains which differ from one another according to the place in the chromosome at which the F-factor is inserted, and marker genes different from the ones shown here can also be used.

The exercise can be organized in various ways in terms of student groupings and the stage at which the procedure is handed over to the class. A convenient approach is to have students grouped into threes and to begin by provisioning each group with a mating mixture in a 500 ml flask, according to the schedule given below. Make sure that sterile technique is used throughout and plates etc. are clearly labelled.

Requirements (per group of three students)
Vortex mixer (but see below),
500 ml flask,
12 tubes of buffer,
36 tubes of soft agar,
Supply of 0.1 ml and 1 ml pipettes,
12 plates of each of the three kinds of minimal media with supplements: i.e. -leu, -pro and -trp.

All of the plates must be supplemented with streptomycin to select against the **Hfr** cells and allow growth of only the F⁻ recombinants. The -leu plates contain proline and tryptophan but not leucine, the -pro plates contain leucine and tryptophan but not proline and the -trp ones contain leucine and proline but not tryptophan.

Method
1 Mix exponentially growing **HfrH** and F⁻ cells in the ratio of 1 : 20, by adding 0.5 ml of the HFr cultures to 10 ml of the F⁻ culture in a culture room or water bath maintained at this temperature.
2 Shake the mixture very slowly for 5 min to mix the cells and to allow contacts to be made between donors and recipients.
3 Withdraw 0.1 ml of the mating mixture and add this to 50 ml of broth in a 500 ml flask (i.e. a 500-fold dilution). Take the time, which becomes time 0. Each

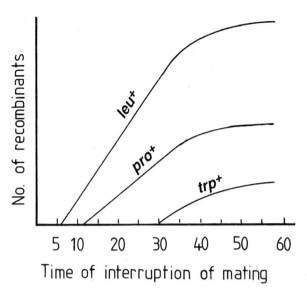

Fig. 10.5 Graph showing the time of entry curves for three marker genes after transfer by conjugation from **HfrH** donor cells in *E. coli*.

group of students is now provided with one flask containing the mating mixture, which is best kept shaking gently at 37°C.

4 At time 0, and at subsequent intervals of +5, 10, 15, 20, 25, 30, 35, 40, 45, 50 and 60 min, withdraw samples of 0.5 ml of the mixture into 2 cm tubes containing 2 ml buffer.

5 'Vortex' each sample for 60s to break apart the conjugating bacteria. If a vortex mixer is not available, separate the conjugating cells by sucking up and blowing out with force several times.

6 To each of three tubes of soft agar add 0.1 ml of the vortexed mixture and shake gently to mix.

7 Pour the soft agar onto the set of three supplemented plates to be used for each sampling time, i.e. onto the -leu, -pro and -trp plates.

For the above steps, a division of labour is useful whereby one person samples the mixture, one person uses the vortex and one pours the soft agar.

8 Incubate the plates at 37°C for 48 h.

9 Count the number of colonies on each of the three plates and record results.

Analysis and interpretation

Plot the numbers of *leu⁺*, *pro⁺* and *trp⁺* recombinants on a graph as numbers of recombinants against time of interruption of mating. Then extrapolate the curves back to the x-axis. The point at which each curve intercepts the x-axis gives the time of entry after mating at which the marker genes enter the recipient F⁻ cells (Fig. 10.5).

It will be seen from Fig. 10.5 that each marker gene has its own curve, and that the number of recombinants increases with the time that was available for mating. Each curve eventually reaches a plateau. The reason for the curves taking this form is that not all of the mating pairs establish contact at the same time, and

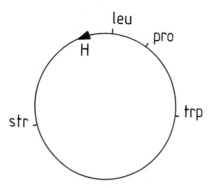

Fig. 10.6 Diagrammatic representation of the chromosome of *E. coli* K12 showing the point of insertion of the sex-factor in strain **HfrH** (H), and the relative positions of the three marker genes used in the exercise on interrupted mating. The arrow at H indicates the place at which the chromosome opens up and its direction of transfer from the **Hfr** donor to the recipient F⁻ cell.

the number of pairs exchanging their DNA increases over the early stages of mating. Later on when all contacts between available pairs have been made the number of recombinants reaches its plateau. Late entering markers plateau at a lower level because they require a longer time period to pass through the conjugation tube and during this time delay many of the conjugation tubes are broken. The time of entry of markers, therefore, reflects their order in the chromosome and their relative proximity to the insertion point H (Fig. 10.6) at which the chromosome opens up to initiate its transfer.

REFERENCES

Clowes R.C. and Hayes W. (Eds.) (1968). *Experiments in Microbial Genetics*. Oxford, Blackwell Scientific Publications.

Sheppard P.M. (Ed.) (1973). *Practical Genetics*. Oxford, Blackwell Scientific Publications.

APPENDIX

INTRODUCTION

This Appendix contains an assemblage of information which is useful for carrying out the procedures and exercises described in this book, but which is not central to their utilization. To avoid obscuring the logical chapter structure we have gathered this accessory material together as a source for reference when needed.

SOURCES OF EXPERIMENTAL MATERIAL

The organisms required for the exercises described in this book can be obtained from various sources. Some materials can be purchased locally, 'home-grown' or collected from wild populations (such as bulbs and seeds for Chapter 1, inflorescences for Chapter 2 and clover specimens for Chapter 7). Other materials can be purchased from Biological Supply Houses, some of which carry a wide range of genetic teaching material. Two main stockists of such material are Philip Harris Biological Ltd., Oldmixon, Weston-Super-Mare, Avon BS24 9BJ, UK and Carolina Biological Supply Company, Burlington, North Carolina, USA. In addition, a local university department will often be more than willing to help out with strains of organisms that they are using routinely for teaching purposes.

It should also be borne in mind that the exercises presented in this book serve as *model* systems, and can be adapted and carried out with alternative materials where these are more readily obtainable.

USE OF THE MICROSCOPE

Chromosomes are small objects that have to be examined with high magnification of a microscope. Microscopes have three lens systems, as follows.

1 Light from the source of illumination first passes through the *condenser lens*, which is beneath the stage. The purpose of this lens is to focus the light onto the specimen on the slide.
2 Light then passes through the specimen and is picked up by the *objective lens*. This lens produces a magnified and more highly resolved image inside the microscope tube.

3 The eyepiece, or *ocular lens*, then magnifies the image in the tube to give the image we see. The total magnification of the final image is, therefore, the magnifying power of the objective lens multiplied by that of the eyepiece. Most microscopes have a set of three objective lenses, namely x10, x40, and x100, together with a x10 eyepiece. These objective lenses, working with the ocular lens, therefore give final magnifications of x100, x400 and x1000, respectively.

In order to see an object clearly the microscope must be working at its optimum efficiency. This can only be achieved by using what is referred to as *critical illumination*. The procedure for setting up the microscope for critical illumination is as follows.

1 Place a slide on the stage and focus on the specimen with the x10 objective lens.
2 Centre the condenser. This is done by closing down the iris diaphragm on the condenser and checking that the circle of light shown appears in the centre of the field of view. If the condenser is off-centre it can be moved by turning its adjusting screws (but note that some condensers are pre-centred during construction of the instrument and thus cannot be adjusted).
3 Wind the condenser up (or down) until the edge of the iris diaphragm comes into sharp focus. At this point the light source is focussed and the microscope is set up for critical illumination. The iris diaphragm is then opened again. It may be used subsequently to adjust the amount of light coming through the tube (do not do this by changing the position of the condenser lens!). Modern microscopes are *parfocal* in that all three objective lenses focus on the same plane. Only minor adjustments with the fine focus are therefore needed when changing from one objective lens to another.
4 The x10 objective lens is now used to search the slide to find the most suitable area of the specimen for study. The x40 lens is then used for more detailed observations and the x100 for the most critical study. When working with the x100 lens *immersion oil* has to be used to fill the air space between the coverslip and the lens itself, to reduce light loss through air. To do this, swing away the x40 lens and place a small drop of oil over the illuminated area of the coverslip before moving the x100 lens into position. The x40 objective lens is normally a *dry* lens and will have to be cleaned should it accidentally become smeared with oil. Ether or xylene are effective solvents for this purpose (though they can damage the cement used to mount the lenses in some microscopes). Check this point with the makers if you are in doubt.

It is advisable for cytological work to use slides of high quality and thin (No.1) coverslips.

SOLUTIONS FOR PRE-TREATMENT, FIXATION, HYDROLYSIS AND STAINING OF CHROMOSOMES

Pre-treatments

Colchicine
Make up a 1.0% stock solution (w/v) by dissolving 1 g of colchicine powder in

100 ml of distilled water. Dilute 1:10 before use with distilled water. Keep both the stock and working solutions in the refrigerator, but remember to bring the solution to room temperature before use. Colchicine is a powder and is usually purchased in 1 g quantities. It is expensive and should be treated as a dangerous chemical. Alternative pre-treatments, as described below, may be more appropriate for school use.

Para-dichlorobenzene (1,4-dichlorobenzene)
This is probably the cheapest and easiest to use of all the spindle inhibitors. It works quite well with plant material. Prepare and use as a saturated aqueous solution (a few crystals in 250 ml distilled water).

α-Bromonaphthalene (1-bromonaphthalene)
Prepare and use as a saturated aqueous solution. This substance is not very soluble in water and 2 or 3 drops of concentrate are enough to saturate 250 ml of water. Place a few drops of 'bromo' in a beaker and fill with water. Make an emulsion by squirting the 'bromo' drops through a Pasteur pipette. Allow any droplets of the 'bromo' to settle and then use the liquid above the droplets for pre-treatment of roots. Do not cover the tube while treating the roots.

8-Hydroxyquinoline
This is used as a 0.002 M solution in water. Add 0.15 g to 500 ml water. Dissolve at 60°C. Allow to cool to 18–20°C before use. Use a fume cupboard when making the solution.

Fixation: acetic alcohol

Use a *freshly prepared* solution made up of 1 part of glacial acetic acid to 3 parts of absolute ethyl alcohol. It is not a good idea to store tubes containing acetic alcohol in a refrigerator, since fumes from the fixative can corrode metal fittings. Fixed material which is to be kept for more than 1 week should be transferred to 70% ethyl alcohol and maintained at about 4°C. Make sure bottles with stored material are well sealed to prevent evaporation of the alcohol.

Hydrolysis: hydrochloric acid

1 M HCl is made by adding 89 ml concentrated HCl to 911 ml distilled water. Always add the acid to the water; *never mix a small quantity of water with a large volume of acid.*

5 M HCl is made by adding 445 ml concentrated HCl to 555 ml distilled water. As above, *add the acid to the water.*

Staining

Feulgen's solution (Schiff's reagent)
Dissolve 0.9 g basic fuchsin in 250 ml 0.15 M HCl, together with 4.8 g sodium metabisulphite ($Na_2S_2O_5$). Leave standing overnight. Then add 1 g activated charcoal, shake vigorously for a few minutes and then filter. Feulgen's solution keeps well

for several weeks in the refrigerator in a dark bottle. It may also be purchased ready-made from usual sources. The 0.15 M HCl is made by adding 3 ml analar HCl to 247 ml distilled water. See the note above concerning adding the acid to the water.

Aceto-carmine
Mix 45 ml glacial acetic acid with 55 ml distilled water. Add 2 g carmine dye and boil gently for 1 h in a reflux condenser. Shake well and filter when cool. Aceto-carmine keeps well, but may need filtering from time to time to remove any precipitate. It may also be purchased in ready-made form.

Aceto-orcein
Mix 2 g orcein dye in 100 ml glacial acetic acid and boil gently for 20 min, ideally in a reflux condenser. Allow to cool and then filter. The mixture has a tendency to come to the boil suddenly and therefore needs handling with care. The solution keeps well, but may need filtering from time to time. Dilute 1 : 1 with distilled water if staining is too intense.

Propionic orcein
Add 2 g orcein powder to 100 ml 45% aqueous propionic acid in a flask. Place the flask on a reflux condenser and boil gently for several hours (*do not allow to boil 'dry' as the semi-solid is semi-explosive*). Cool the solution and filter it into stock bottles. For use, dilute 1:1 with 45% propionic acid and filter. Ideally, filter each time before use.

Snow's alcoholic HCl carmine
Mix 4 g carmine dye in 15 ml distilled water to which 1 ml concentrated HCl has been added. (See note above concerning mixing water and acid.) Boil gently for 5 min. Allow to cool. Then add 95 ml 95% ethyl alcohol and filter. The solution lasts indefinitely and only occasionally needs further filtering.

For acidified 70% alcohol (used for holding stained material prior to slide preparation), add 10 ml concentrated HCl to 90 ml 70% alcohol and mix.

MAKING SLIDES PERMANENT

Temporary preparations of cells in mitosis or meiosis, in which the coverslip is sealed with rubber solution or nail varnish, will last only a few days. To make a temporary preparation permanent it is necessary to remove the coverslip, dehydrate the material on the slide and then embed the material in a mountant under a new coverslip.

1 *Removal of the coverslip*. The ideal way to do this is to freeze the whole slide in liquid nitrogen and then prise the coverslip off, as follows. Hold the slide with a pair of long-handled tweezers and immerse it in a thermos flask of liquid nitrogen for about 15 s. After bubbling has ceased lift the slide out and place it coverslip uppermost on the bench. *Without delay*, hold one edge of the coverslip with the fingers, using a piece of blotting paper as protection, and then

quickly prise off the coverslip by inserting a razor blade (single edged) under the opposite edge. The bulk of the material will remain on the slide. The coverslip is usually discarded.

2 *Dehydration*. Place the slide immediately into 95% ethyl alcohol for 1 min, and then into absolute ethyl alcohol for a further minute. Coplin jars are ideal containers for doing this; it is important to remember which side of the slide carries the material!

3 *Mounting*. Transfer the slide to a 1:1 eucalyptus oil/absolute ethyl alcohol mixture for 1 min, then to 100% eucalyptus oil for a further minute. Remove and drain the slide and then add a small drop of euparal, or some other appropriate mountant over the material. Apply a new coverslip. Flatten the coverslip with gentle pressure and allow a few days for the mountant to harden before using the slide.

An alternative method for removing the coverslip is to float it off in a shallow dish of 45% acetic acid. The slide is then transferred to 75% alcohol, followed by the same procedure as in (2) above. With this method some material is often lost unless an adhesive such as glycerine albumin has been applied to the slide before making the preparation.

PURIFYING AND REVITALIZING STOCK CULTURES OF DROSOPHILA

Contaminated Cultures

Stock cultures of Drosophila can occasionally become contaminated. The main culprits are mites, yeast and other fungi, and bacteria. For badly contaminated cultures it is best to obtain a new stock from a supply house. But a mildly contaminated culture can be 'rescued' as follows.

Prepare and label a series of four or five new containers of food, paying special attention to sterility. Transfer to bottle 1 a few flies from the contaminated culture. For mite contamination it is best to select 'mite free' (or nearly so) flies under a dissecting microscope. After one day or so transfer these *same* flies to bottle 2. Allow one or two days for egg laying to occur and then transfer the flies to bottle 3, and so on. Finally discard the flies. The aim is to rid the flies of mite eggs, fungal spores, etc. by contact during the serial transfers. Bottles 4 and 5 should then be incubated and the eggs reared to adulthood, after which there will be a strong probability that the new cultures will not be contaminated.

Poorly Performing Cultures

Stock flies become strongly inbred (by brother/sister matings, for example) and thus may lose vitality through the accumulation of deleterious alleles in homozygous condition. Such flies may be difficult to maintain as stocks and may perform poorly in an experiment. To revitalize the stock it is necessary to take a sample of flies through a mating with flies of a different line (usually a wild type line), followed by re-isolating the required strain from the F_2. The procedure to adopt is simply an extension of the one that is used to demonstrate monohybrid or dihybrid inheritance in a class exercise, the only extra requirement being to isolate female flies of the desired phenotype from the F_2 before they have mated. These flies are then used with corresponding males to re-establish a new culture. This new culture will carry many new alleles from the outcrossing and will thus be revitalized.

CHI-SQUARED TESTS

For Analysis of Genetic Ratios

In genetical experiments in which we study ratios of progeny phenotypes, such as those of an intercross or testcross, the results that we obtain are subject to *sampling error* and thus usually do not fit exactly with the expected ratios. For instance, a Drosophila male heterozygous for a pair of alleles at a particular locus (say vg^+/vg) will segregate the two kinds of alleles in exactly equal numbers into its sperms, because meiosis in each spermatocyte will give two vg^+ and two vg carrying sperms. But when mating takes place only a *small sample* of sperms will actually function as gametes, and it is a matter of chance whether any particular sperm carries the vg^+ or the vg allele. Chance events are likewise involved in the eggs that are fertilized. Because of these chance events in heredity the proportions of phenotypes in our samples of F_2 or testcross progeny do not conform exactly with the expected theoretical ratios. There is nearly always a certain *deviation* between the observed and expected values due to sampling error. The smaller the samples we work with the greater this sampling error is likely to be. The principle is just the same as that involved with tossing a coin and observing the numbers of heads versus tails in relation to the number of tosses made.

In order to assess the magnitude of the deviation between our observed result and that expected on the basis of a particular hypothesis, such as a 3:1 ratio, we use the *chi-squared* (χ^2) test. This simple statistical procedure takes account of the size of the sample while assessing the deviation between the observed and expected numbers, and gives us an answer as a single numerical value. We then use this χ^2 value to determine the probability that the deviation between observed and expected can be accounted for by chance, or whether there is a real reason for the departure from the expected ratio.

The formula for calculating χ^2 is:

$$\chi^2 = \Sigma \frac{(O-E)^2}{E},$$

where Σ = sum of, O = observed value, E = expected value. The test is conducted as follows.

Suppose the observed values in a Drosophila monohybrid cross for normal x vestigial wing were 390 normal wing and 114 vestigial wing, out of a sample of 504 F_2 flies. The expected values for a 3:1 ratio are 378 and 126 respectively, i.e. $\frac{3}{4}$ of 504 and $\frac{1}{4}$ of 504. The calculations we need are shown in Table A.1.

We now consult a table of the distribution of c2 (Table A.2) to find out what the probability is of obtaining a deviation as large as, or larger than, the one we have by chance alone. Table A.2 takes account of the fact that the size of the χ^2

Table A.1 Calculations for a chi-squared test.

Phenotypic class	O	E	O-E (= d)	d^2/E
Long wing	390	378	+12	0.38
Vestigial	114	126	− 12	1.14
Total	504	504	0	$\chi^2 = 1.52$

depends upon the number of independent comparisons (phenotypic classes) which are added up to give its value, as well as the size of the deviations in the individual classes. We allow for the number of independent comparisons in our test by using the *degrees of freedom (d.f.)*, which appear down the left-hand side of Table A.2.

Table A.2 Distribution of χ^2.

Degrees of freedom	Probability								
	0.99	0.95	0.80	0.70	0.50	0.30	0.20	0.05	0.01
1	0.00016	0.004	0.064	0.148	0.455	1.074	1.642	3.841	6.635
2	0.0201	0.103	0.446	0.713	1.386	2.408	3.219	5.991	9.210
3	0.115	0.352	1.005	1.424	2.366	3.665	4.642	7.815	11.345

(From Fisher and Yates: *Statistical Tables for Biological, Agricultural and Medical Research*, published by Oliver and Boyd. Reproduced by permission of the authors and publishers.)

Two classes have only 1 degree of freedom, because there is only one independent comparison, i.e. given the number in one class the number in the other class is fixed for a constant sample size. The number of degrees of freedom for the test described here, and in the example that follows, is *one less than the number of classes*. In the present example, therefore, we have 1 degree of freedom and we use the top line of Table A.2. We see that our χ^2 value of 1.52 is less than 1.642 for $P = 0.20$ (or 20%), but more than 1.074 for $P = 0.30$ (30%). This tells us that if there were no real deviation from a 3:1 ratio, and we repeated the experiment a large number of times, deviations as great, or greater than, the one we have here (+ or - 12) would occur in more than 20% of the cases due to chance alone. In other words, there is a high probability that the deviations which we obtained in our experiment are due to chance alone, and we can confidently conclude that there is *no significant departure* from a 3:1 ratio, i.e. our results agree with expectation. The level of probability at which we decide to accept or reject our hypothesis of agreement with a given ratio is usually taken as 5% ($P = 0.05$). Any value of χ^2 smaller than 3.841, for 1 degree of freedom, is considered to be non-significant and the observed values are then said to agree with expectation. A value of χ^2 larger than 3.841, for 1 degree of freedom, is considered to be 'significant at the 5% level of probability'. We then say that the probability that the deviations are due to chance is very low (less than 1 in 20), and there must be some real departure from the expected ratio. It is then another matter to decide upon the possible reasons for the lack of agreement.

The chi-squared test can be used to test the significance of any ratio in a genetical experiment. As another example, consider Mendel's results for the testcross experiment involving the character pairs round, wrinkled seeds versus yellow, green seeds in peas. The heterozygote (R/r,Y/y) was testcrossed to the double recessive (r/r,y/y) and gave the four phenotypic classes shown in Table A.3. The hypothesis to be tested is for agreement with a 1:1:1:1 ratio. (Mendel himself was not able to make these tests because the statistical procedures for doing so had not been worked out at that time.)

Table A.3   Chi-squared test on Mendel's testcross results for the dihybrid cross involving two seed characters in peas.

Phenotypic class		O	E	d	d^2/E
round yellow	(RY)	55	52	+3	0.173
round green	(Ry)	51	52	- 1	0.019
wrinkled yellow	(rY)	49	52	- 3	0.173
wrinked green	(ry)	53	52	+1	0.019
Total		208	208	0	$\chi^2 = 0.384$

Since there are four phenotypic classes in our test we use the row in Table A.2 that corresponds to 3 degrees of freedom. Our χ^2 value falls between probability levels of $P = 0.80$ (or 80%) and $P = 0.95$ (or 95%), and is therefore non-significant. The probability that the observed deviation from a 1:1:1:1 ratio is due to chance is greater than 80%, i.e. there is no real departure of the observed from the expected values. The data agree with the hypothesis.

Consumer warning
1 Especially note that a chi-squared test must be conducted on *actual numbers* of things, *not* on their proportions or percentages or their ratio. This is because percentages, for example, do not take into account the size of the sample of objects from which the percentages are calculated. If we are dealing with proportions or percentages or a ratio at any stage and then wish to conduct a chi-squared test, we must first convert these to actual numbers of things (for which of, course, we need to know the sample size).
2 The chi-squared test can lead to erroneous conclusions when one or more of the observed categories is small (5 or less). There are mathematical ways of correcting for this effect (Yates' correction, for example). The interested reader should consult a text on statistics for further information. The main point of interest here concerns the importance of having large sample sizes when dealing with genetic data, especially when one or more observed categories is likely to be small in number.

Contingency χ^2

In some situations we may wish to compare our results not against a particular ratio but rather with another similar set of data. For example, we may wish to compare the numbers of female and male Drosophila flies homozygous for *cu* that have curled, semi-curled and not-curled (straight) wings, to see if males and females are significantly different from each other in this respect. We can use the chi-squared test to do this, but since we are dealing with samples for which we do not have an overall set of expected values we have to use what is known as a *contingency (or heterogeneity) chi-squared test*. To do this we set up our results as shown in Table A.4.

Table A.4 Calculations for a contingency chi-squared test.

	Curled	*Semi-curled*	*Not-curled*	*Total*
Female	149 (132)	64 (72)	20 (29)	233
Male	124 (141)	85 (77)	41 (32)	250
Total	273	149	61	483

In parentheses in Table A.4 are given the values we expect if the male and female data are simply independent samples from a homogeneous population in which males and females do not differ in their wing curliness. Thus, for example, we expect 233/483 x 273 female curled wing flies; and 250/483 x 61 male not-curled flies. The expected values have been rounded to the nearest whole numbers (for greater accuracy one or two decimal places should be used). They add up to the observed totals all ways, thereby showing us that the calculations have been done correctly.

Our χ^2 value is then calculated exactly as before, by summing $(O-E)^2/E$ for each pair of values in the table. However, the calculation is more easily done by summing the values O^2/E and then subtracting the grand total. Thus we have

$$\chi^2 = (149^2/132 + 64^2/72 + 20^2/29 + 124^2/141 + 85^2/77 + 41^2/32) - 483$$
$$= 494.3 - 483$$
$$= 11.3$$

The degrees of freedom are given as the number of rows minus 1 multiplied by the number of columns minus 1, i.e. (2-1) x (3-1) = 2. From Table A.2 we see that with 2 degrees of freedom the probability of getting a χ^2 value of 11.3 or greater by chance alone is less than 0.01 (1%). We conclude, therefore, that there is a significant (real) difference between males and females as to their wing curliness.

A contingency chi-squared test can be used with any number of rows or columns (more than one each, of course). The calculations for χ^2 and for the degrees of freedom are done simply as an extension of the example described above.

SAFE HANDLING OF ORGANISMS, ESPECIALLY MICRO-ORGANISMS

The organisms used in the exercises described in this book are all considered to be completely safe for handling by human beings. It must be emphasized, however, that the culture media and the vessels used for growing micro-organisms such as yeast and bacteria are susceptible to accidental contamination by other organisms which may be pathogenic, and this possible hazard must be allowed for in the safety procedures that are used.

Possible Modes of Infection by Micro-organisms

In the handling of culture materials and vessels it is possible to inhale microbes either as dry spores or on water droplets during accidental spillage. By the proper use of sterile techniques pathogenic strains can be excluded, so that even if such accidents do occur there is no health hazard. Accidental spillages or breakages can also bring organisms into direct contact with the skin, and antiseptic swabs should then be used for cleaning. The use of teat pipettes eliminates the possibility of sucking any dangerous organisms into the mouth. Immediate and proper disposal of broken glass greatly reduces the chance of taking organisms (and chemicals) into the body via accidental cuts.

Rules of Safety

In addition to taking the sensible precautions outlined above it is good practice to observe the following general safety rules in the laboratory, whatever material is being handled.

1   Wear a protective laboratory coat.
2   Wear solid shoes; wearing sandals or going bare footed in a laboratory is hazardous.
3   Do not eat, drink or smoke in the laboratory.
4   Protect cuts and grazes with a waterproof dressing.
5   Swab spillages immediately.
6   Always use sterile techniques when working with micro-organisms.
7   Avoid culturing or sub-culturing unknown organisms, particularly those coming from human sources.
8   Autoclave used pipettes and culture vessels at 15 p.s.i. for 20 min to kill all living organisms.
9   Seal cultures with adhesive tape during incubation, and keep them sealed for examination. When petri dishes have to be opened for sub-culturing or for access to fruiting bodies, lift the lid carefully and keep it off for the minimum time necessary to perform the required operation.
10   Ventilate the laboratory well, especially when using ether.
11   Keep stock bottles of ether in a fume cupboard (hood).
12   Be especially careful with a naked bunsen or spirit lamp flame. Do not use ether anywhere near such a naked flame.
13   Wash hands thoroughly before leaving the laboratory, especially prior to eating.

Sterile Technique

'Sterile technique' means working with fungi and bacteria in such a way that the organisms themselves, and the media on which they are growing, are contained within a sterile environment (e.g. a screw-cap bottle or a petri dish); and that during handling they are protected from outside contamination and are themselves prevented from contaminating the outside environment or the operator.

To maintain these sterile conditions when working on an open bench, be sure to select a draught-free area and to swab down the work surface with 70% alcohol before use.

Sterilize an inoculation loop by heating it to red-hot in a bunsen flame, right back to the handle. A few seconds cooling will be needed before the loop is ready to use. Sterilize glass spreaders by flaming after dipping in 95% alcohol, repeating the procedure two or three times.

To keep culture bottles sterile while opening, first loosen the cap and then remove it in the crook of the little finger of one hand while holding the bottle with the other hand. Flame the neck of the bottle using a bunsen or a spirit lamp, and then introduce or remove the specimen with a sterile loop. Reflame the neck of the bottle and screw back the cap as quickly as possible.

Plastic petri dishes are gamma irradiated after packaging and are sterile when purchased. Keep them within the pack until needed, and then only take out the required number and reseal the pack. When medium is being poured into a petri dish, or organisms introduced, the lid should be only partially removed to gain access, and kept open for as short a time as possible.

REFERENCES

Anon (1981). Safety in school microbiology. *Education in Science*, **92**, 19–27.
Imperial College, University of London (1974). *Precautions Against Biological Hazards*.
Mather, K. (1951). *Statistical Analysis in Biology*. 4th edition. London, Methuen and Co.
Philip Harris Biological Co. (1981). *Safe Handling of Micro-organisms*.
Sharma A.K. and Sharma A. (1965). *Chromosome Techniques. Theory and Practice*. London, Butterworths.

GLOSSARY OF TERMS

Acrocentric (chromosome). A chromosome whose centromere is close to one end, thereby producing chromosome arms of very different lengths.

Agglutinate. Clump, as in red blood cells treated with ABO antigens.

Aleurone. The outermost nutritive tissue of kernels of maize and other grasses.

Allele (= allelomorph). One of two or more forms of a gene. Alleles occupy corresponding loci on homologous chromosomes.

Anaphase. A stage of mitosis and meiosis in which chromatids or half-bivalents completely separate from each other and move to opposite poles of the spindle. **Anaphase I (II).** Anaphase of the first (second) division of meiosis.

Anthocyanin pigment. Red pigment found in some plants such as maize.

Antibody. A specific substance produced in response to the introduction of an antigen into an organism, for the purpose of eliminating (neutralizing the effect of) that antigen.

Antigen. A substance, living or dead, which causes the production of an antibody when introduced into an organism.

Ascogenous hyphae. Filaments (hyphae) from which asci arise in an ascomycete fungus.

Ascomycete. One of the main groups of fungi, characterized by the production of spores (**ascospores**) by meiosis within an elongate sac (**ascus**).

Ascospores. See under **ascomycete**.

Ascus (-i). See under **ascomycete**.

Ascus initial cell. A cell destined to produce one or more asci in an ascomycete fungus.

Asexual reproduction. Development of new individuals without sex; reproduction based on mitosis rather than meiosis and fertilization.

Autosome. Any chromosome other than a sex chromosome; thus **autosomal linkage**, etc.

Axillary bud. See under **terminal bud**.

Balancing selection. Natural selection in which two (or more) opposing forms (genotypes) are favoured in their survival and reproduction, so that both forms survive in balance in a population.

Basic number. See under **x**.

Binucleate. Having two nuclei.

Bivalent. An associated pair of homologous chromosomes formed by synapsis during the early stages of meiosis. Bivalents separate into **half-bivalents** during the first division of meiosis.

C-mitosis. Colchicine mitosis, in which the drug colchicine (or equivalent agent) has inhibited spindle formation, thereby allowing the chromosomes to be well separated from each other during slide preparation. Their prolonged stay in prophase also makes the chromosomes shorter than normal.

Cell cycle. The life cycle of a cell, consisting of a non-dividing stage (in which genes are active and in which DNA and chromosomes replicate) and a dividing stage (mitosis).

Centromere. The region of a chromosome (chromatid) at which the spindle is attached and by whose activity a chromosome moves during prometaphase and anaphase.

Chiasma (-ata). That region(s) of a bivalent at which the pairing partners of chromatids change, as a result of crossing over (**chiasma formation**) by breakage and rejoining of non-sister chromatids.

Chromatid. One of a pair of longitudinal subunits of a replicated chromosome, which separate from each other during anaphase of mitosis and anaphase II of meiosis.

Chromatin. The material of chromosomes of an interphase (non-dividing) nucleus, consisting of DNA and associated protein and other substances.

Chromosome. Elongate structure of the nucleus in which the genes are contained. Chromosomes are especially visible when the nucleus divides.

Chromosome abnormality. An atypical number or structure of the chromosomes of an organism.

Chromosome arm. One of two sections of a chromosome separated by the centromere. **Arms lengths** of a chromosome are therefore determined by position of the centromere, which may vary in different chromosomes.

Chromosome complement. The set of chromosomes characteristic of an organism; see also under **karyotype**.

Chromosome contraction (condensation). The process during prophase of mitosis and meiosis in which chromosomes gradually become shorter and thicker and thus more clearly visible. This process is reversed during telophase.

Chromosome doubling. Chromosome replication (chromatid formation) but without an ensuing mitosis. The chromatids separate from each other 'passively', and remain in the same nucleus.

Cleistothecium (-ia). As for perithecium but without an opening; spores are liberated by breakdown of the wall of the fruiting body.

Codominant. Condition in a heterozygote in which the effects of both alleles are manifest in the phenotype.

Cohesive force. A force or structure of unknown nature that binds sister chromatids together until they begin to separate at the start of anaphase.

Colony. A group of cells (or organisms) that are usually the products of asexual reproduction (mitosis) of a single parent cell or organism.

Complementation (genetic). The phenomenon in which two genomes, each mutant in a different way, can produce a normal phenotype when they are together in the same cell.

Congression (of chromosomes). The movement of chromosomes into the equator during prometaphase of mitosis and meiosis. See also under **prometaphase**.

Conidiophore. A mycelial structure that produces (carries) conidia.

Conidium (-a). A spore(s) produced during asexual reproduction, especially in fungi.

Conjugate mitosis. The simultaneous mitotic division of two nuclei.

Conjugation (bacterial). The transfer of genetic material from a donor to a recipient organism by direct cell contact. cf. **transformation, transduction.** Thus **conjugation tube** (the region of contact between two bacterial cells, allowing for conjugation).

Crossing over. The process(es) leading to the recombination of linked genes, especially by chiasma formation.

Cytokinesis. Division of the cytoplasm of a cell by cleavage (in an animal cell) or cell plate formation (in a plant cell). Division of the nucleus and of the cytoplasm together constitute division of the cell.

Cytoplasm. The entire contents of a cell except its nucleus and (in plant cells) the cell wall, bounded by a membrane (the plasmalemma).

Diakinesis. The final stage of prophase I of meiosis in which chromosome contraction is more-or-less at its maximum and in which chromatids may rotate at chiasmata to give 'cross-road' appearances.

Dihybrid. Organism formed from mating parents that differ in respect of two pairs of alleles (contrasting traits); thus **dihybrid cross** (the cross involved), **dihybrid ratio** (the ratio of F_2 progeny types of a dihybrid cross).

Diploid. 2x, in which there are two of each type of chromosome in a cell or organism.

Diplotene. Stage of prophase I of meiosis in which homologous chromosomes separate from each other except at chiasmata and in which contraction of chromosomes is well advanced.

DNA replication. The process whereby the gene material DNA (deoxyribonucleic acid) is faithfully reproduced.

Dominant (-ce). A character or its corresponding gene (allele) which is manifest fully in the phenotype when in either homozygous or heterozygous condition.

Double crossover. The occurrence of two crossing over events (chiasmata) between two given gene loci, involving the same or different chromatids.

Egg. A female gamete.

Endosperm. A nutritive tissue surrounding the embryo in many seeds.

Enzyme. A biological catalyst.

Enzyme deficiency complementation. See under **complementation** (genetic).

Epistasis. A form of interaction between non-allelic genes whereby the effect of one hides the presence of another in the phenotype.

Equator. The plane of the spindle at right angles to and usually midway between the two spindle poles. The equator is usually located more-or-less in the centre of the cell.

Eukaryote. An organism whose genetic material is separated from the cytoplasm by a membrane, thus forming a nucleus.

Expressivity. The degree to which the effect of a gene is shown in the phenotype of **penetrant** individuals.

F_1. The first filial generation (offspring) produced from mating two parents.

F_2. The second filial generation (offspring) of two parents, produced by intercrossing or self-fertilizing the F_1 generation.

F^-, F^+. The sex factor(s) of the bacterium, *Escherichia coli*.

Feedback inhibition. A process whereby accumulation of the product of a biochemical pathway inhibits further production of this product, usually at an early point in the pathway.

Fertilization. The process of fusion of gametes to produce a zygote during sexual reproduction.

First division segregation. Segregation (separation) of alleles during the first division of meiosis, which results when there is no chiasma formation (crossing over) between the centromere and the gene concerned.

Fitness (physiological, reproductive). The survival or reproductive potential of an organism or genotype, compared to others or compared to the mean of a population.

Fixation. The killing and preserving of biological material in as near life-like a state as possible; thus **fixation** (an agent that fixes material).

Fruiting body. A well-defined group of fungal mycelia and the spores they carry and surround.

Gamete. A cell (or nucleus) that fuses with another during fertilization to produce a zygote.

Gametic (number of chromosomes). See **n**.

Gene. A unit of heredity that determines directly or indirectly a phenotypic trait, consisting of a particular linear sequence of nucleotides of DNA.

Gene conversion. The process by which small sections of DNA are converted from one sequence of nucleotides to another, usually during recombination of linked genes in a heterozygote.

Gene frequency. The proportion of a particular gene (allele) relative to the total number of alleles at a given locus in a breeding population; thus also **genotype frequency**.

Genetic drift. Random as opposed to directed changes in gene frequencies in a population (usually a small one) generated by chance reproduction or survival of forms not representative of the whole population.

Genetic engineering. The production of novel combinations of DNA (see under **recombinant DNA technology**) and their incorporation into a host organism.

Genetic map. See under **linkage** (map).

Genetic material. The substance comprising the genes of an organism, i.e. DNA.

Genetics. The study of heredity.

Genome. The set of chromosomes and genes of an organism.

Genotype. The composition of an organism in respect of the individual or the collective genes it carries.

Genotype frequency. The proportion of a particular genotype relative to others in a population.

Half-bivalent. The products of separation of a bivalent during the first division of meiosis, consisting of a pair of chromatids (which, however, are not necessarily genetically identical, because of crossing over).

Haploid. See under **x**.

Hardy-Weinberg equilibrium (principle, law). A mathematical expression of the genotypic composition of a population, which remains constant (in equilibrium) in the absence of such factors as mutation, selection and non-random breeding. For a pair of alleles **A** and **a**, the equilibrium is expressed as p^2 (**AA**) + $2pq$ (**Aa**) + q^2 (**aa**), where p and q are the frequencies of the two alleles in the population.

Hemizygous (-ote). The state of (organism) possessing only one allele at a given locus, rather than the usual two alleles, usually in relation to certain genes in the sex chromosomes.

Heredity. The biological phenomenon by which parents and their offspring are similar and dissimilar as a result of the genes they carry and pass on.

Heritability (h^2). That proportion of the variance of the phenotype that is genetic in origin.

Heterokaryon. A fungal mycelium containing two genetically different nuclei; cf. **homokaryon**.

Heteromorphic (chromosomes). Chromosomes that are different in form, in respect of length, centromere position, etc. Sex chromosomes are usually heteromorphic.

Heterothallic. Able to undergo sexual reproduction only with another one of different sexual behaviour.

Heterozygous (-ote). The state of (organism) possessing different alleles at one or more gene loci; cf. **homozygous (-ote)**.

Hfr. High frequency recombination; a bacterial strain in which the sex factor is joined to the main bacterial chromosome, thereby giving the organism high fertility.

Homokaryon. A fungal mycelium containing two genetically identical nuclei; cf. **heterokaryon**.

Homologue (-ous) (chromosomes). Chromosomes that are identical to each other in respect of their morphology (length, centromere position, etc.) and in their genetic loci. Homologous chromosome pair up and segregate from each other during meiosis.

Homothallic. Able to undergo sexual reproduction (mating) with itself or with another mycelium of the same type as far as its sexual behaviour is concerned; cf. **heterothallic**.

Homozygote (-ote). The state of (organism) possessing identical alleles at one or more gene loci; cf. **heterozygous (-ote)**.

Hypha (-ae). One (many) of the filaments that comprise the body (=thallus) of a fungus.

Inbred line. A group of related individuals that are more or less genetically homozygous as a result of systematic matings between siblings or between siblings and their parents.

Incomplete (-ly) dominance (-t). A genetic situation in which the phenotype of the heterozygote is intermediate to the phenotypes of the two corresponding homozygotes.

Independent assortment. The random segregation of different pairs of alleles located on different (non-homologous) chromosomes, as a result of random chromosomes segregation of these chromosomes during meiosis.

Interacting genes. Genes whose action combine to produce a phenotypic trait; epistasis.

Interphase. The stage of the cell cycle in which chromosomes are not distinctly visible and in which gene activity and DNA and chromosome replication take place.

Karyotype. The set of chromosomes characteristic of a particular individual or species, etc., defined in terms of chromosome number, relative lengths and centromere positions.

Kernel. The embryo and its surrounding nutritive tissues and skin in maize.

Leptotene. The earliest stage of prophase I of meiosis in which chromosomes appear very long and thin.

Linkage (partial, complete). The tendency of certain genes to be associated in their inheritance, as a result of their location in the same chromosome. Thus **linkage group** (those genes that show linkage). The number of linkage groups equals the number of different (non-homologous) chromosomes of an organism. Also **linkage map** (the linear sequence and relative distances apart of the genes in a chromosome, based on their percentages of recombinant progeny).

Locus (-i). The site (place) of a gene in a chromosome or in its linkage map.

Map unit. The unit distance between linked genes on a linkage map, corresponding to 1% recombinant progeny.

Mapping (genetic). The process of arranging gene loci as to their linear sequence and distances apart, based on the percentage of recombinant progeny they produce.

Mating type. The character of an organism or cell as to its sexual behaviour.

Meiosis. A pair of nuclear divisions in which the zygotic number of chromosomes is reduced to the gametic number and in which the genetic material is shuffled (recombined). The cells formed from meiosis are either gametes or spores.

Metacentric (chromosome). A chromosome whose centromere is more or less in the centre, so that the two chromosome arms are approximately equal in length; thus also **submetacentric.**

Metaphase. The stage of mitosis and meiosis in which the chromosomes have reached their final attachment to the spindle and orientation in the equator of the cell. Thus **metaphase I (II)** (metaphase of the first (second) division of meiosis). cf. **prometaphase.**

Micro-evolution. Small scale evolutionary changes at the level of the species and below, often expressed as changes in gene frequencies.

Mitosis. The division of the nucleus of a cell to give daughter cells which have the same chromosome number and the same genetic make-up; = **karyokinesis.**

Mitotic cycle. The cell cycle from prophase to telophase, but not including interphase; = **mitosis.**

Monohybrid. Organism formed from mating parents that differ in respect of a single pair of alleles (contrasting traits); thus **monohybrid cross** (the cross involved); **monohybrid ratio** (the ratio of progeny types of a monohybrid cross). Thus **dihybrid, trihybrid,** etc.

Multinucleate. Containing two or more nuclei.

Multiple alleles. A group of more than two alleles of a locus (gene), of which a diploid organism can carry any two.

Mutagen. See **mutation.**

Mutation. A heritable change in the genetic material (other than that brought about by recombination); thus **mutagen** (an agent that produces a mutation); **mutant** (an organism carrying a mutation or the altered gene itself).

Mycelium. The collection of hyphae making up the body of a fungus.

n. The number of chromosomes in a gamete of an organism; the gametic number.

2n. The number of chromosomes in the zygote (and thus somatic cells) of an organism; the zygotic number.

Natural selection. Selection occurring naturally (rather than being of human origin) in which certain forms (genotypes) are favoured in their survival and reproduction at the expense of other forms.

Non-allelic. Not belonging to the same set of alleles.

Non-homologous (chromosomes). Different in terms of gene loci. Non-homologous chromosomes do not pair up (synapse) during meiosis.

Non-recombinant. See under **recombinant.**

Non-sister chromatids. Chromatids (usually of a bivalent) belonging to different, non-homologous chromosomes; cf **sister chromatids.**

Nuclear envelope (= nuclear membrane). A double membrane surrounding the nucleus of a cell. The membrane disappears at the end of prophase in mitosis and meiosis and reforms during telophase.

Nucleolus (-i). An organelle(s) of the nucleus representing the site of synthesis of ribosomes.

Nucleolus organiser. The region on one particular chromosome or chromosomes at which the nucleolus is formed during telophase and from which the nucleolus disappears at the end of prophase of mitosis and meiosis.

Nucleus. The organelle of a cell containing the bulk of its genetic material (DNA).

Orientation (of chromosomes). The process by which chromosomes connect at their centromeres to the spindle, the poles of which they therefore face.

P. The parental generation of a cross.

Pachytene. The stage of prophase I of meiosis in which homologous chromosomes have completely paired along their lengths.

Parthenogenesis. Development of a female gamete (into an embryo) without it first having been fertilized by a male gamete.

Penetrance. The proportion of individuals that show in their phenotype the effect of a particular gene they carry.

Perithecium. A flask-shaped fruiting body of an ascomycete fungus, consisting of a mycelial wall and opening and containing the asci.

Phenotype. The individual or collective visible properties of an organism, including morphological, biochemical and behavioural aspects.

Plasmid. A small, circular piece of DNA present within the cytoplasm, independent of the main chromosome, in a bacterium.

Pleitropy (-ism). The production of multiple, apparently unrelated effects of a single gene (allele).

Pollen mother cell. A cell that is destined to undergo meiosis to produce pollen in the anther of a plant.

Polygene (-ic). One of many genes which have small and usually additive effects on a particular character, usually a quantitative character.

Polymorphism (genetic). The existence of two or more forms (genotypes) of a species within the same population at frequencies that cannot be explained solely by mutation.

Population (Mendelian or breeding). The individuals of a species at a given locality that are potentially able to breed with each other.

Progeny test. A genetic cross, such as a testcross, set up to determine the genotype of the progeny or parents of an individual.

Prokaryote. An organism whose genetic material is not separated from the cytoplasm by a membrane; cf. **eukaryote.**

Prometaphase. The stage of mitosis and meiosis following prophase in which chromosomes become attached to the developing spindle and move into and become oriented in the equator of the cell. Thus **prometaphase I (II).** Prometaphase of the first (second) division of meiosis.

Prophase. The stage of mitosis and meiosis in which replicated chromosomes are transformed into condensed and clearly visible structures prior to their attachment to the spindle.Thus **prophase I (II).** Prophase of the first (second) division of meiosis. Prophase I is often sub-divided into **leptotene, zygotene, pachytene, diplotene** and **diakinesis.**

Qualitative (character). A trait in which only a few alternative and distinctive states occur, which are therefore usually written descriptively (qualitatively); cf. **quantitative character.**

Quantitative (character). A trait such as height which shows a wide spectrum of alternatives, which are therefore described numerically.

Recessive. A character or its corresponding gene (allele) which is manifest in the phenotype only when in homozygous condition; cf. **dominant.**

Reciprocal cross. One of a pair of crosses in which the sexes are reversed; thus **a** x **A** and **A** x **a.**

Recombinant. An individual or cell or set of genes that results from rearrangement of genetic information of male and female genomes; thus **non-recombinant** (=parental) combination of genes or organisms.

Recombinant DNA technology. The artificial (human) cutting and splicing of DNA from different sources to generate novel combinations.

Recombination (genetic). The rearrangement (shuffling) of genetic information of male and female genomes of an organism, as a result of different relative orientations of chromosomes and chiasma formation during meiosis.

Replication (DNA, chromosomes). The process whereby DNA and chromosome material is accurately copied to give more of its kind. For a chromosome the products of replication are called **chromatids.**

Saprophyte (-ic). An organism that obtains its food from dead organic (animal or plant) matter.

Second division segregation. Segregation (separation) of alleles during the second division of meiosis, which results when there is crossing over (chiasma formation) between the centromere and the gene concerned.

Segregation (genetic). The separation of pairs of alleles into different cells, as a result of separation of homologous chromosomes during meiosis.

Selection. Differential survivability or reproductive potential of different organisms or genotypes relative to others or relative to the mean of a population.

Sex chromosome. A chromosome involved in the determination of sex (male or female) in an organism; thus **X chromosome** and **Y chromosome**.

Sex factor (= fertility factor). A small piece of DNA (plasmid) responsible for determining mating type in bacteria.

Sex linked. Genes or their phenotypic effects whose pattern of inheritance is linked with the sex of the organism concerned; thus **sex linkage** (the association of genes in the sex chromosomes).

Sexual reproduction. Reproduction involving sex, i.e. meiosis and fertilization.

Sister chromatids. The products of replication of a chromosome. Sister chromatids separate from each other during anaphase of mitosis.

Somatic cells. Cells of the body of an organism, other than those associated with reproduction.

Sperm. A male reproductive gamete.

Spermatocyte. A cell that is destined to undergo meiosis to produce sperm in the testis of an animal.

Spindle. A barrel-shaped structure composed of microtubules and other substances that is formed when a cell divides by mitosis or meiosis, and to which the chromosomes become attached via their centromeres for movement during prometaphase and anaphase. Thus **spindle fibres** (the fibrous, microtubular elements of the spindle).

Spindle pole. One of two regions of a dividing cell at which spindle fibres focus and to which chromosomes move during anaphase.

Synapse (-is). The longitudinal pairing of homologous chromosomes during early stages of meiosis.

Telophase. The final stage of mitosis or one of the divisions of meiosis in which, after reaching the poles in anaphase, the chromosomes return to their uncondensed (interphase) state. Thus **telophase I (II)**. Telophase of the first (second) division of meiosis.

Terminal bud. The growing point and surrounding young leaves at the tip of a shoot; thus **axillary bud** (found in the angle formed between a stem and a leaf or between two stems).

Testcross. A cross of an individual, known or hypothesized to be heterozygous at one or more loci, with an individual that is homozygous recessive for the allele(s) concerned. Thus **testcross ratio** (the ratio of individuals (genotypes) produced in a testcross).

Tetraploid. 4x, in which there is four of each type of chromosome in a cell or organism concerned.

Trait (genetic). A character or aspect of the phenotype.

Transduction. The transfer of DNA (genetic material) from a donor to a recipient organism or cell by a virus.

Transformation. The uptake and incorporation of naked DNA by a recipient organism or cell such that its genotype is changed.

Translocation (chromosomal). A chromosome abnormality in which segments of chromosomes have changed their position by chromosome breakage and rejoining (other than in chiasma formation).

Trihybrid. See under **monohybrid**.

Triploid. 3x, an organism or cell in which there are three of each type of chromosome.

Trisomy(-ic). Condition in which there is one extra chromosome in an otherwise normal and diploid set.

True breeding. Producing progeny of the same type as itself.

Variance. A statistical measure of variation. Thus V_P (variation of the phenotype); V_G (variation of the phenotype that derives from genetic components (genes) of an individual); V_E (variation of the phenotype that derives from environmental components).

Wild type. The predominant type (strain, organism or gene) in a population.

x. The basic and haploid number of chromosomes, in which there is only one of each type of chromosome.

X chromosome. See under **sex chromosome**.

Y chromosome. See under **sex chromosome**.

Zygotene. The stage of prophase I of meiosis in which pairing of homologous chromosomes takes place.

Zygotic (number of chromosomes). See under **2n**.

REFERENCES

Abercrombie M., Hickman C.J. and Johnson M.L. (1962). *A Dictionary of Biology*. London, Penguin.

Rieder R., Michaelis A. and Green M.M. (1976). *Glossary of Genetics and Cytogenetics*, 4th edition. Berlin, Springer-Verlag.

INDEX